D1092814

CURIOUS MINDS

CURIOUS MINDS

The Power of Connection

PERRY ZURN AND DANI S. BASSETT

The MIT Press
Cambridge, Massachusetts
London, England

Portions of chapter 4 were published in "Busybody, Hunter, Dancer: Three Historical Models of Curiosity," in *Toward New Philosophical Explorations of the Desire to Know: Just Curious about Curiosity*, ed. Marianna Papastephanou (Cambridge: Cambridge Scholars Press, 2019), 26–49. They are published with the permission of Cambridge Scholars Publishing.

The MIT Press would like to thank the anonymous peer reviewers who provided comments on drafts of this book. The generous work of academic experts is essential for establishing the authority and quality of our publications. We acknowledge with gratitude the contributions of these otherwise uncredited readers.

This book was set in Adobe Garamond Pro by New Best-set Typesetters Ltd. Printed and bound in the United States of America.

Library of Congress Cataloging-in-Publication Data

Names: Zurn, Perry, 1981- author. | Bassett, Dani S., author.
Title: Curious minds : the power of connection / Perry Zurn and Dani S. Bassett.
Description: Cambridge : The MIT Press, 2022. | Includes bibliographical references and
  index.
Identifiers: LCCN 2021049885 | ISBN 9780262047036 (hardcover)
Subjects: LCSH: Curiosity. | Thought and thinking. | Neurosciences.
Classification: LCC BF323.C8 Z687 2022 | DDC 153.4/2—dc23/eng/20211216
LC record available at https://lccn.loc.gov/2021049885

10 9 8 7 6 5 4 3 2 1

To all the children who question whether it needs to be this way.

Contents

Preface: Our Story

There are so many silences to be broken.
—Audre Lorde, *Sister Outsider* (1984)

In a sense, this book represents the thought of one mind and two bodies. It is like us, identical twins, a sketch upon the canvas of life drawn by one DNA and two experiences. It was penned from one amniotic sac and two breaths. It is the product of one upbringing seen through the eyes of two genders. Both one and two, both I and we, both same and different, the origination of this book reflects the central tensions of curiosity's nature. Curiosity is one word, one string of letters, one concept in the mind, but curiosity also has multiple manifestations, a plethora of practices, and kindred kinds in many bodies. Like a genus spanning species or our mother embracing her eleven children, curiosity is both one and many. The one-and-many nature of curiosity is an opportunity for the attainment of epistemic freedoms: we are permitted to be curious! But it is also a liability for the perpetration of epistemic injustices: we are permitted to be curious in less than many ways.

As children sitting beside one another at identical desks, we breathed the air of a uniquely curious space embedded in familial relations. Our schoolroom was adjacent to the living room and beside the kitchen in a stone ranch house in Central Pennsylvania. Surrounded by more than a thousand books and guided by our mother's one mantra, to love learning, we saw the beauty of what could be known. "Take control of your education," she would say, "no one is spoon-feeding you. Figure out what you want to know and then figure out how to learn it." As the family grew in size, our mother quickly taught us to teach ourselves, to grade our own work using the homeschooler's teaching manual and solution key, and to rework problems until we figured out why we had gotten them wrong. We were encouraged to choose the order and nature of subjects each day, each month, each year. A full semester on bones? No problem. Devising experiments with poisonous plants from the nearby woods? Sure. Practicing Johannes Brahms's Rhapsody in B minor op. 79, no. 1 for eleven hours per day? Why not? Centering history, science, language, and art curricula all around mushrooms? A definitive "yes." Our mother's goal was to follow our interests, connect concepts, fuel our questions, and open doors, situating our inquiries in divergent historical trajectories and knowledge communities. In concrete ways, we were given the fodder for a curiosity of epistemic freedom and accountability.

Yet in our beautiful little curious world, freedom was circumscribed, in time, body, and conceptual space. Girls were meant to grow up into women, get married, have babies, care for the home, and be submissive and obedient to their husbands; college and career were not in the cards. Boys were meant to grow up into men, acquire a trade, provide financially for their families, and rule the household, choosing their wife's friends and arranging their daughters' marriages. With twin eyes, we looked at each other, then out at our world, and asked, "Why? You gave us spades. We plan to dig." Buoyed by the kindness of unlooked-for allies along the way, we broke down walls, crossed boundaries, and scaled heights to become the interdisciplinary scholars that we are today—scholars who are committed to recognizing and resisting the epistemic inequities that surround and suffuse us. We are definitively not what we were meant to be, but we are following the ever-becoming trajectory of the curiosity instilled in us: one that spans, one that connects, one that embraces, and one that builds; one that appreciates the crosscurrents and coalitions within and through which we come to know. "The real enemy of the human race is not the fearless and irresponsible thinker, be he right or wrong," writes educator Abraham Flexner. "The real enemy is the man who tries to mold the human spirit so that it will not dare to spread its wings."[1] While our local world was just such an enemy, the curiosity our mother instilled in us was an antidote, a counterpoint, a friend.

Some would see our story as somewhat surprising, certainly strange, and perhaps downright alien. Yet many others would view our story as fairly commonplace and fully expected. The uncanny familiarity of our story stems from an uncomfortable fact: curiosity is policed everywhere. The demarcation of curiosity for some bodies and not others, some times and not others, some spaces and not others is not sequestered to small islands of human life but rather pervades all of human society. People of differently assigned genders are given gifts and experiences to encourage their curiosity in some directions and not others. Doors are opened for people of some racial or ethnic backgrounds, classes, sexualities, and abilities and not others, in some cultures and not others. While the disadvantaged are held back, the privileged are offered freedom for their minds to roam the fields, scale the summits, and meander the byways of thought in search of knowledge. Even in institutions of higher learning, which we might expect to be the places where the human spirit can spread its wings, scholars are hired most often when their work falls into a prespecified space: that of a department, a siloed discipline, or a reductively stipulated history or methodology of thought. Where does the narrowing, the constraining, the ruling of curiosity come from? Why do there exist supposedly right and wrong ways to be curious? Worthy and unworthy topics of curiosity? The naturally curious and the naturally incurious?

In part, the answer lies in the oneness of curiosity. *Curiosity is one word, one string of letters, one concept in the mind.* To make one is to draw a boundary, a border, a wall that separates the one from anything else. When you have one, you also have not-one. The existence of presence marks the existence of absence. When you have something, then you must also have nothing. Unity permits dichotomy. Unity chooses simplicity over complexity. In using a single word, a single concept, a single history, we allow curiosity to be defined, parameterized, and, yes, eventually regulated. What people have perhaps too often forgotten is the manyness of curiosity: *its multiple manifestations, its plethora of practices, and its kindred kinds in many bodies.* Despite the one-and-many nature of curiosity, many people hearken only to the former and eschew the latter. But need they? What kind of world might embrace curiosity's manyness? How might we define a multitudinous curiosity? What disciplines might we need to span? What relations might we grasp? What humilities and histories might we need to embrace? What architectures of the mind might we build? How might we consciously practice curiosity in a manner that allows each human spirit to fly? In asking these questions (and many sibling questions), we probe in this book the central tension of curiosity's nature. Borrowing poet T. S. Eliot's words of love to frame our pursuit of curiosity, we "lift and drop" each "question on your plate." We take the time "for a hundred indecisions, and for a hundred visions and revisions." We take the time to wonder, "Do I dare?" And as we do, we find one simple answer: curiosity does dare. It dares "disturb the universe."[2]

A Brief Note on Visual Media

The book's words are complemented by two types of visual media, each chosen to highlight distinct facets of curiosity's mode of being. First, we precede each chapter with a frontispiece created by artist and illustrator Poonam Mistry, emphasizing key concepts in masterfully crafted visual form. Poonam's style incorporates her love of nature, and explores the relationships between patterns, shapes, and colors, creating beautifully intricate illustrations. Her clients include the Commonwealth Education Trust, Amnesty International, American Express, BP Oil, Michael O'Mara Books, Buster Books, Lantana Publishing, Tate Publishing, and Penguin Random House. Well-crafted illustrations that are shaped slowly over long periods serve to foreground the timeless nature of curiosity, and the expanse of questions that burn within us for weeks, months, years, and decades. Such illustrations also foreground the curiosity that leads us to build expertise and to hone a craft.

Second, we pepper several of the more scientific chapters with sketches hand drawn by one of the authors and depicting the concepts as if we were sitting at the pub together, scribbling away on a paper napkin. For scientists, sketches are a crucial part of how we think, how we question, and how we communicate our questionings to others. We try not to leave the house without a pen or pencil, and you'll frequently find us scribbling on stray bits of paper or paper products, including napkins, paper towels, receipts, parking tickets, and book margins as well as on blackboards and whiteboards, or in the sand or dirt with a bit of a stick. Humanists with a book, a sketchpad, or a whiteboard are similarly sketch- and scribble-happy. This type of visual media is therefore anthropologically appropriate to scholars and thinkers everywhere, but it is more than that. Sketches bring curiosity to the here and now, to the fleeting moment of time, to the dashing off ideas before we lose them in the dark recesses of our minds. Sketches also bring curiosity to the deeply personal: to me drawing for you, or you drawing for me, so that we may share in this curious space together.

INTRODUCTION: A NEW VANTAGE POINT

Bugs. Sometimes, something just bugs you.[1] A worm in your ear. We have all had the experience. You are going about your day, and something prompts you to wonder. You mull it over. You try on this explanation or that one. And then you get distracted and you move on. Or maybe you don't, and the worm digs in deeper. Or maybe you do, but the worm returns to its wriggling later that night. You can't shake it off. Maybe you pull out your phone or strike a few keys—or turn to a colleague or ping a friend. Gosh, now you really want to know! Perhaps you hit a few walls—paywalls or prejudices, differences of opinion or the limits of science, or even congenial scoffing at your pet project. Depending on who you are, you might also encounter outright sexism or racism, classism or ableism—all ways of telling you that your bug is a bust. Forget this fleeting interest and focus on something that matters. Regardless, let's say, you carry on. Maybe the kids are screaming, or you are in a meeting, or you drive to the store for milk. Despite bombastic blasts from every corner, you hold onto what's bugging you, refusing to let it bugger off. And then it happens . . . your mind begins to dance and to weave. Collecting the bits of things that might be relevant and stitching them together. So builds the briefest of webs. Perhaps it was a silkworm after all! This is your brain on curiosity.

And you are in good company. Rewind back to 1928, if you will. Author Virginia Woolf is sitting on the banks of a river (the Thames, perhaps, or the Cam, or the Isis) and she has a worm in her ear: women and fiction. What even are they? she wonders. "Questions," she calls them, "unsolved problems." She finds herself walking and thinking, calmly at first and then feverishly. Crisscrossing the Brontë sisters and George Eliot, Charles Lamb and William Thackeray, she wonders not only about women and about fiction but also about their relation and the several ways in which it can be characterized. She strides past the nearby university, a place filled—as she puts it—with obsolete old minds, bereft of body and free of fact, loosed from the roughshod rambling toward truth that she is currently undertaking.[2] And then there it is!

> Thought—to call it by a prouder name than it deserved—had let its line down into the stream. It swayed, minute after minute, hither and thither, among the reflections and the weeds, letting the water lift it and sink it, until—you know the little tug—the sudden conglomeration of an idea at the end of one's line.[3]

This is curious thought. Holding out a line. Casting and recasting it. Waiting patiently, watching the river, the waves, the light, sinking into the circle of life. Until suddenly you feel it: a series of nibbles, then tugs, and then slowly something coalesces there on the end of your line. Lines of thought agglutinate. They weave themselves together and take shape, gaining felt sense and even power. Aha!

But the challenge of it all is not simply patience. It is also the conundrums and the curtailments. What sort of fish did she find? Was it an as yet undiscovered and unnamed specimen? Or was it a species common to the area, such as a carp perhaps or a perch? Was it a mutant fish, strange from the start? Or was it a chimera of a fish, some fin-bearing aquatic creature that only medievalists can fully appreciate? Would taxonomies need to bend for it or break? Or would they simply breeze by it? What in fact was it she discovered? What inkling took shape for her in an instant? We will never know, for she was quickly accosted by an old-fashioned version of the campus police. She had wandered onto the college green, where women were forbidden. "He was a Beadle; I was a woman. This was the turf; there was the path." She rights herself forthwith and heads for the library, only to be informed at the door that she needs a man to escort her in. Alas! She is out walking alone. In both cases, she laments, "They had sent my little fish into hiding." Determined, she returns to casting her line, over and over again, until the arc of her queries weaves a tapestry: *A Room of One's Own*. In this masterful piece, she puts two and two together: thinking requires the freedom to fish, or "a woman must have money and a room of her own if she is to write fiction."[4]

In *Curious Minds*, we start right there, with the architectures of (and for) curious thought itself. How does curiosity move, and how does it build? What are the neural, epistemic, and social bases of that building? How does your brain connect disparate ideas and string together concepts? How does it reconfigure old notions and weave together new ones? Do different people practice curiosity in different ways, building knowledge configurations of different shapes and styles? And how exactly do we practice curiosity together? What does it mean to track down shared hunches? Or to cross-pollinate curiosities across cultures? Is it even possible to characterize how curiosity feels and functions on these multiple registers? And what happens when curiosity falters? By what neural substrates, social structures, or cultural strictures does curiosity get curtailed? For that matter, what would it mean to facilitate the dynamic force of curiosity equitably, cultivating greater freedom of inquiry and counteracting long-standing patterns of epistemic hierarchy? These and so many other questions fuel the book in front of you.

For too long—and still too often—curiosity has been oversimplified. In everyday parlance, it is frequently reduced to the simple act of raising a hand or voicing a question, especially from behind a desk or a podium. Or to turning an object this way, then that, or perhaps simply "googling" something. Scholars generally boil it down to "information-seeking" behavior or a "desire to know." But curiosity is more than a feeling and certainly more than an act. And curiosity is always more than a single move or a single question. From the toddler to the most well-trained scholars, it is rarely if ever a straight shot. Curiosity comes in waves—mutating, agglomerating, washing ashore, and washing back out again. Each wave breaks by spilling, plunging, surging, or collapsing. Curiosity is as much the brisk steps in search of a new vantage point as it is the silent daydreaming by a river's edge. What falls out of most quotidian—and even scholarly—characterizations of it are the lines that make curiosity what it is. The wandering tracks, the weaving concepts, the knitting of ideas, and the thatching of knowledge systems. In a word, the very essence of curious movement in conceptual and physical space. What gets lost are the networks, the relations, between ideas and between people.

In this book, we offer a new theory of curiosity. Drawing on our own fields of complex systems and neuroscience, literature and philosophy, we propose a network account of curiosity.[5] For us, curiosity is a network practice, a relational practice. It works by linking ideas, facts, perceptions, sensations, and data points together. Yet it also works within human grids of friendship, society, and culture. Rambling over vast interdisciplinary terrain—as we are wont to do—we develop an account that is as relevant to the humanities and the sciences as it is to everyday life. When you query, you connect. And you thread not simply one piece and another together but also within a constantly changing web of individual and collective knowledge. Importantly, that curious connection is never just free play. As Woolf so deftly dramatizes, we weave our knowledge webs within existing contexts rife with constraints—constraints not only in the epistemic architectures of knowledge but in the social architectures of knowers too. Here in these pages, we tackle that complexity. Curiosity *connects*. And it does so within the *connective tissues* of brain and body, system and society.

And *that matters*.

CURIOSITY ACROSS THE AGES

Curiosity is as old as the hills, and the study of curiosity itself has a long and storied history. Given its presumed affinity for all things intellect, curiosity has typically been divorced from the body—and from society for that matter—and located instead in the individual organ of thought. But of course that organ is quite slippery, having been understood differently across the vagaries of time. Thousands of years ago, it was believed that curiosity lived in the seat of the soul. Following the medieval period and the Renaissance, curiosity slowly took up residence in the mind. Then with the development of early psychology and neuroscience, curiosity has most recently been located somewhere in the brain—with its precise coordinates still up for debate. Regardless of where curiosity is situated, its consistent restriction to the mental apparatus has been accompanied by a steady divorce from material bodies and community relations. Curiosity has been isolated. But the tide is turning. The deeper we head into the brain, the more impossible that isolation is to sustain. The mass of humanity exerts an unquestionable gravitational force pulling us outward. Reckoning with the network substrate of the brain sends us scrambling back out to account for the many other networks within which curiosity works. As the tides turn, our waves of inquiry pull sand and seaweed away from shore. Appreciating the multiple architectures that make thought possible—whether neural, knowledge, or social networks—is in fact crucial not only to account for curiosity but also to practice it intentionally.

Saint Augustine is credited with perhaps the earliest and most definitive account of curiosity in the Western canon. A well-educated North African rhetorician in Numidia, Augustine converted to Christianity at the age of thirty-one and went on to become a church bishop, writing definitive treatises in the faith. For Augustine, curiosity resides in the soul.[6] Sometimes he characterizes it as a "lust of the eyes," or a cupidinous interest to see and to know things of earthly but not eternal import. At other times he depicts it as a form of "lower reason," which turns the light of the soul dim with fleshly desire and delight in secular learning. Whatever the specific formulation, Augustine understands curiosity as a formidable force that moves the soul away from God and Christian fellowship, drawing it ever deeper into the epistemic intrigues of a mutable, corruptible world. As such, it is a "disease," he says, that disrupts the order of creation and wrenches an individual human from their proper place in it. Now, while we have long relinquished the conviction that curiosity is a sin, we have held onto a sense that curiosity lives in the soul, or at least in that part of us where we wonder, learn, doubt, and remember. Augustine's account is, then, a direct predecessor of psychological and neuroscientific accounts of curiosity today.

But there is an intermediary step. As the Roman Catholic empire waned, early modern science arose and, with it, the unabashed celebration of all the elements of scientific inquiry: observation and hypotheses, yes, but also curiosity. Against all the religious vitriol curiosity had sustained, Francis Bacon, onetime lord chancellor of England and staunch proponent of empiricism, defends it, characterizing it as absolutely "natural" and in fact "useful" when tasked with building knowledge in the service of humankind.[7] Similarly, René Descartes, French mathematician and cornerstone of modern philosophy, naturalizes the intellectual impetus to inquire and to explore. Curiosity and wonder, he says, are both located in the mind, where "animal spirits" (forces carried in the blood) hold up different perceptual impressions before the mind's eye. In the case of wonder, the spirits are staid and calm; in the case of curiosity, they are a bit jittery and quick. Through the sibling emotion (or as Descartes has it, the "passion") of generosity, curiosity and wonder can fuel the organism's acquisition of knowledge beneficial to society as a whole.[8] Lifting curiosity from the soul and to the mind, then, the early modern period directly links it to the biologically rooted inner eye. From there, the brain is not far behind.

In contemporary psychology and neuroscience, curiosity is located in the brain and, depending on the study, in specific parts of the brain. This localization makes good sense given that these fields currently understand curiosity to be an information-seeking behavior that is driven to fill information gaps. Curiosity would then naturally be located in the information processing center: the brain.[9] But where in the brain? Psychologist Daniel Berlyne locates curiosity in the interplay between the reticular arousal system and "neural assemblies" in the cerebral cortex.[10] In the subsequent work of neuroscientist Jacqueline Gottlieb, data scientist Min Jeong Kang, and psychologist Celeste Kidd, among others, curiosity has migrated across gray matter tracks depending upon the object of study. When curiosity is associated with exploration, it appears in the frontopolar cortex, intraparietal sulcus, or posterior cingulate cortex; with novelty, it appears in the ventral striatum; with trivia, in the caudate nucleus and inferior frontal gyrus; with learning and memory, in the hippocampus, dopaminergic regions, and nucleus accumbens; or with attention and decision-making, in the frontoparietal system.[11] And while the nervous system extends throughout the body, and is tuned by environmental and social factors, today's science rightly insists that the brain—and its contribution to curiosity—is importantly unique.

The deeper into the brain you press, however, the farther back out you have to pivot. There, in fold after fold of gelatinous tissue, are masses of cell nuclei, axons, and synapses, shuttling electrical signals this way and that in a riotous web of neural activity. Far from chaotic, those neurons work in clusters, yes, but also across clusters. And these clusters swing and even dance. They do so, moreover, in ways partially determined by genetic and epigenetic substrates, environmental influences (both present and past), and social interactions. This is a complex system if there ever was one. A veritable symphony. To its credit, network science—building on sociology, art theory, and philosophy from the 1960s and 1970s—provides a renewed invitation to attend to these networks in the brain, but also to the knowledge networks and social networks within which that complex neural system functions. The field summons us to appreciate the network architectures not only of the thinking organ but also of thought itself as well as thinkers of today and yesteryear. In this book, we heed that call.

And none too soon. In the epic march from soul to mind to brain, we have lost touch with some things along the way. We have lost track of the networks. We have largely forgotten to wonder how curiosity builds or breaks community, whether with other humans or with a more-than-human reality. We have largely forgotten to ask about curiosity's relationship to emotions, to trauma, and to other affective bodily phenomena. We have hardly wondered where else in the body curiosity is lodged. And we have left largely unexplored curiosity's relationship to bodily movement—or kinesthetics—as a whole. We believe so heartily that curiosity is crucial to societal achievement, moreover, that we have spent little energy in cultivating its attunement with societal generosity. And we have barely paused to ask just what links are activated (or lie dormant) between curiosity and culture, social values, and even politics.[12]

There is so much left to investigate! Foregrounding curiosity as connectional allows us to turn renewed attention to the brain as a complex system. It prompts us to ask what insights other complex systems—such as language or music—might provide in the project of probing the curious mind. It also attunes our attentions to the epistemic and social systems that enable the brain to function in the first place. It prompts us to ask about the organ of learning itself, *and* simultaneously about the architecture of what is learned and the arrangement of learners in the process. To really understand curiosity, after all, it matters not simply with what and how one learns but with whom and where as well. As such, the network account we propose invites ever-richer circuits of analysis, opening up a whole new world behind and ahead of us.

FOOTPATHS AND FLIGHT WAYS

In *Curious Minds*, we trace a path through this network of questions. In no way do we imagine that this is the only, or indeed the most direct, course. We would be disappointed if it were. Rather, we hope the path takes on a life of its own. Sometimes we hue closely to what we know and at other times we stretch out to touch what we do not yet know. We try our footing, we reroute our flight. We take our coordinates from the depth of the humanities and the height of the sciences by turns. In so doing, we marvel at and blithely explore the vast landscape that a network approach opens up for the study of curiosity.

In chapter 1, "The Science of Curiosity," we begin by noting the centrality of curiosity to the very spirit and practice of science. There *is* no science without curiosity. We then turn to the contemporary science of curiosity, focusing on the recent contributions of psychology and neuroscience. Studies in psychology utilize quantitative and qualitative measurements to understand the function of curiosity, while studies in neuroscience make use of brain-imaging techniques (e.g., functional magnetic resonance imaging [fMRI]) to identify the neural substrates of curiosity. Whatever the methods, each field has made immense strides in illuminating the behavioral and neural components of information seeking. There remains, however, an important limitation to this work. It functions largely on the information gap theory, according to which curiosity is an intrinsic motivation to fill a knowledge gap. Within an information-seeking framework, curiosity remains acquisitional, a project of acquiring—and amassing—information. But this is not curiosity's only function.

In chapter 2, "Curiosity as Edgework," we identify the roots of the acquisitional model of curiosity in the longtime predecessor of psychology and neuroscience: philosophy. Traditionally, philosophy characterizes curiosity as the desire or the appetite to know something. Inspired by network theory, however, we pivot from the paradigm of curiosity as nodal acquisition to curiosity as edgework. We propose a new framework that understands curiosity as a practice of connection, a way of building relations between and among systems of knowers, knowledges, and things known. To develop this framework, we revisit the history of philosophy and excavate a different genealogy for curiosity—one that emphasizes not informational units but relations instead. We find precedent in both Western and Indigenous lineages for this conceptualization of curiosity as a connector rather than conquistador. Finally, we reflect on the immensely promising lines of study that this view poses for psychology and neuroscience today.

In chapter 3, "The Network Paradigm," we develop the theory of curiosity as edgework through the terminology and methodological resources of network science proper. Specifically, we suggest defining curiosity as a practice of knowledge network building. Drawing on concepts and tools from network science, neuroscience, psychology, and linguistics, we posit that knowledge is a network and that it can in fact be mathematically represented as such. Moreover, if knowledge is a network, and curiosity is the growth principle of that network, then it is theoretically possible to quantitatively characterize and mathematically model that growth using tools from network science. That is, we may be able to graph how ideas slowly accrete into architectures of individual and collective knowledge. Our proposal here formalizes many of the intuitions we have about curiosity as relational, and by that formalization it provides the foundations from which to construct testable hypotheses for future empirical work.

In chapter 4, "Curiosity's Got Style," we ask the next logical series of questions. If curiosity is a network building activity, are there different architectures and styles? The equivalent of an Antoni Gaudí and a Frank

Lloyd Wright, a Beyoncé and a Nina Simone? And might these differences vary with neural idiosyncrasies, personalities, sociocultural values, and more? In short, we answer "yes." Diving back into the heart of Western intellectual history, and out past its edges, we identify three styles of curiosity that are relatively consistent across time: the busybody, the hunter, and the dancer. In their curious trajectories, the busybody collects informational tidbits, while the hunter barrels down on one particular interest, and the dancer takes creative leaps of imagination. Each has a distinct kinesthetic signature, creating loose knowledge networks, tight networks, and loopy ones, respectively. In ideational and social spheres, moreover, each style has a unique function, mapping out different routes through knowledge grids and cutting paths across social networks.

In chapter 5, "Webs of Knowledge," we ask, But are these paths traceable in empirical research? Are styles of building observable in laboratory settings? They are indeed. We report on our study of knowledge network building among users of *Wikipedia*, where participants spanned from busybodies (who surfed among loosely connected *Wikipedia* pages) to hunters (who surfed among closely connected pages). Yet this is only the beginning of future work. If people have different styles of knowledge network building, how might those differences affect education or other learning environments and activities? We know that human brains learn clustered or modular networks better than random or disordered ones. Might those clusters be clustered in different ways for different knowers? And how would styles of clustering even be achieved in a learning environment where information is taught linearly through books, lectures, and lesson plans? Ultimately we ask, How might the vast variety of learners' interests and capacities be best facilitated in and beyond the classroom?

In chapter 6, "Curiosity Takes a Walk," we pause for a moment. In network science, the path by which one traverses a network is called a "walk," and different walks on a network conform to different constraints. Likewise, the path by which one traverses a *knowledge* network can be a walk within specific constraints. The suggestion that walking is a way of thinking—a way of negotiating or generating knowledge—has long roots in the humanistic tradition, as does its converse suggestion: that thinking is a way of walking. Here we investigate four sorts of walks, which themselves involve specific forms of curious thought: the philosophical walk, the spiritual walk, the environmental walk, and the political walk. These ways of moving through physical space map onto ways of moving through conceptual space. We therefore propose that one critical way to cinch together a philosophy and network science of curiosity is precisely by thinking these geographies and graphologies of inquiry together.

In chapter 7, "Your Brain on Curiosity," we offer our most systematic and experimental account of the network neuroscience of curiosity. Because, as we argue, curiosity is not solely about the acquisition of a piece of knowledge but also about the edgework of connecting pieces of knowledge, we explore not only the relations (and networks) that make up the brain but also the principles by which they may be encoded in the brain, how they are learned and remembered. Drawing as freely from neuroscience as from poetry, music, and even lexicography, we investigate what actually happens in your brain on curiosity. What correspondences does and might the architecture of your brain have with the architectures of the knowledge it produces and processes? Canvasing the function of brain regions and their interrelations, we emphasize the necessary flexibility between them. We then speculate on the current trajectory of neuroscientific inquiry into the neural basis of knowledge network building.

In our final chapter, "Reimagining Education," we take a step back to appreciate the big picture. If curiosity is best understood as a practice of knowledge network building, of connecting and relating informational bits and informational knowers, what should education look like? How can we reimagine learning differently, more energetically, more equitably? To answer this question, we propose renewed attention to the three key networks involved: neural, social, and knowledge networks. Recognizing the immense diversity that exists within each of these network substructures, we sketch out the paths to a more inclusive, dynamic, and transdisciplinary pedagogy. In that inquiry, we dig deeply into the case studies of neurodiverse writer Naoki Higashida, the civil rights movement's Freedom Schools, and interdisciplinary scholar par excellence Leonardo da Vinci. Committed to reducing the social inequalities and loosening the traditional disciplinary boundaries that subtend the academy, we aim to facilitate freedom for *everyone*'s curiosity to pursue seemingly incommensurable ideas and applications.

Curious Minds closes with our own ruminations on the future of curiosity. Anchoring our reflection in our interdisciplinary and social justice commitments, we share what we take to be the next steps in the emancipation of inquiry, the signposts on our journey. As we imagine greater freedom for the dynamism and diversity of curiosity, we think not only of the networks yet to be built but also of the networks that need to crack in order to make room for new building. We close, then, with a rumination on cracks and their role in crafting a more equitable world for all curious minds.

OSCILLATIONS

The account we offer here, as with many ideas intertwining minds and fields, has been developed in the patient, sometimes haphazard space of creativity. It has both lurched and lightly danced across time—five years to be exact. The academic life provides all kinds of opportunities for thinking together, whether giving talks or lectures, writing essays, conducting experiments, jumping on Zoom calls, or catching a weekend Islay Scotch (neat). We swapped many a midnight text and daylight delight as this project took shape. For us, it all began with a flight of fancy. What if we were to produce a bit of scholarly work together? Where might our respective areas of expertise meet? In what ways could our fields inform one another? From philosophy to neuroscience, from literature to physics, we noticed that together we revel in ancient and modern ways of thinking about thinking. How, why, what, with whom, and where do we think? On what grounds do we declare, exclaim, or question? As humans, we engage so dramatically with our inner and outer worlds, like soft fuzzy bumblebees wriggling, rambling, and waggling on the miniature flower meadows of spirea bushes. Absolutely enthralled. Completely enmeshed. What is this enchantment that draws us out, draws us in, and calls us to question? Curiosity. Curiosity was the kind of concept that could travel well across the vast space within and between our fields. But curiosity how? We started to sleuth. What do we do as academics, what do we love as humans, and what does that tell us about curiosity, about how we practice it, but also about how we might theorize it? *Curious Minds* is the record of our journey.

What you have before you, then, is truly a collective effort. The ideas developed here could not have germinated or borne fruit without us thinking together—that is, without our relationship. We practiced curiosity collectively, connecting philosophy and neuroscience, literature and psychology, social justice and network science, so as to build a specific idea, with its very own idiosyncratic knowledge network architecture. That said, for the purposes of the work, we had distinct authorial responsibilities. Dani holds the primary authorship of chapters 1, 3, 5, and 7, while Perry holds the primary authorship of chapters 2, 4, 6, and 8. We have done nothing to mask the distinctiveness of our writing and thinking styles. Rather, we invite readers to come along on our journey. This book is not a monologue but rather a dialogue. And it is not, as such, the most straightforward of books; it does not follow well-trimmed paths. Instead, this is a dynamic investigation, a live-wire query. We gather disparate pieces together, dive down rabbit holes, and leap across vast swaths of cleanly curated space, just to see where we land. And we do that purposively.

But there is also a clear method to the maelstrom, in both form and content. As we toggle between our voices, our fields, and the cadence of our respective walks, we craft an inquiry that expands and contracts as it goes. We call up different canons, explore different peripheries, and take scholastic risks of different altitudes depending on our field-specific expectations. We do this because we believe in the promise of curiosity, but also the pleasure of curiosity. We stretch physics to music and poetry, while we stretch critical theory to graphology and neurology. And there is a flexibility in our partnership here, a little swing dance. You will spot Perry ambling among the archives and see Dani standing still, staring at brain images in the laboratory. But you will also notice Perry sitting down to taxonomize and watch Dani running off to tell stories. What better way to follow the call of curiosity and to court serendipity, though, than to weave between disciplines and ways of knowing, defying borders and boundaries as we go?

As a medium of consilience, we cultivate a literary style, courting an attunement to sound and rhetorical form in a way that enhances our arguments and better equips our explorations. There is a depth of wisdom in language's flexibility, in the soma of its poetry. In step with that literary quality, moreover, we quote other writers generously, mesmerized by the genius not only of what they say but of how they say it. As thought artists, we work in mixed media, assembling mental material from the twigs and seeds, dyes and putties of language, written across times in various genres and by distinct voices. Implicit here in the work of writers, scholars, and artists centuries in the making are the elemental insights of a network account of curiosity. Furthermore, interweaving the voices of others throughout our journey is no mere ploy of stylistic enrichment or evidentiary justification. Our conceptualization of curiosity as relational necessitates that we draw from and build upon multiple knowers and ways of knowing, whether cradled within our own disciplines and the academy or out past their edge.

There is nevertheless a distinctive anchor for each of us in this inquiry. We are a trained philosopher and physicist, respectively, working in our own fields of critical social theory and network neuroscience. As such, we toggle throughout this book between the mental and the social, between brains and other structures of belonging. In doing so, we behave a bit like mycologists.[13] After all, mushrooms typically steal the show, with their big shiny heads, but beneath and around them there stretches mile upon mile of mycelia, a whole underground fungal community. Similarly, the brain typically takes the credit for curiosity, but it functions within a vast interconnective system that makes curiosity possible. In *Curious Minds*, we shuttle back and forth between the mushrooms and mycelia, the brain and the social body, in order to track curiosity *where it lives*, across neural and social networks.

We invite our readers to come along on this journey, to take a walk with us. To hold the thread of our argument, but also to revel in the riffs and digressions. The book performs the curiosity of which it speaks. As such, we invite readers not only to appreciate but also to *participate*—to be curious *with* the book, not simply about it. Curiosity engages the unknown and relishes in the process of coming to know; it is that path (and not solely the path's destination) that we celebrate in our writing. We encourage readers to wonder with us, to toggle between the disciplinary resonances, and to dance in this newly choreographed space. What we offer here is an experience as much as an idea. And we do so with the conviction that a robust theory is more powerful than any prescriptive "how-to" manual. Through it, readers of all sorts can reimagine the practice of inquiry in innumerable ways in their everyday lives.

WHYS AND WHEREFORES

Imagine that curiosity is less like stockpiling treasures than swinging from branches on the tree of life. That tree is one of many trees that make up a forest of knowledge. This forest is forever growing and changing based on the configuration of the soil, the directionality of the sun, and even the age of the earth. These natural constraints on the growth of the knowledge forest, moreover, are not alone. You, alongside many other human and nonhuman creatures, cultivate the forest, shepherding its growth, distributing its decay, and appropriating portions for your use. That too is curiosity. Curiosity is a growth principle in a forest ecology of knowers and knowns, would-be knowers and would-be knowns. And that ecology shifts and shapes depending on prior growth patterns as well as power struggles and generosities between growers. There is a delightful chance and happenstance to the roots curiosity threads underground and to the branches curiosity weaves in a canopy overhead. But there are also societal logics and neural constraints that often determine which idea trees get nurtured best and celebrated most, at what times and in which places.

Curiosity, after all, like worms and webs, fish and forests, is situated within a larger ecology. What then are the relevant distribution patterns, relationships, and interdependencies within which curious thought functions? Throughout this book, we aspire to ecological validity by studying curiosity in situ. Noticing it *where it lives*, specifically where it lives *between things*. Between competing knowledge networks, between old nodes and new, and between knowledge network builders. Curiosity lives on the edge.

It walks a line. As such, it is impossible to formally isolate. It cannot be reduced to simple, singular acts, nor can it be unified and relieved of all its cracks. In reimagining curiosity, we bring neuroscience and psychology, philosophy and literature, to converge on the real story: curiosity as a network practice, a relational practice.

And why does this matter? It matters in part simply because this is our curious minds at work—for a moment, in a delightful act of freedom, we have found a space to breathe. More fundamentally, it matters because not all curious minds are free. Nor are all curiosities given the space to breathe. Woolf's frustrations still ring painfully in our ears. And she is far from alone. Ta-Nehisi Coates, reflecting on his own Black boyhood in Baltimore, regrets that schools exacted "compliance" more than they encouraged "curiosity."[14] Similar testaments abound, showing the ways in which societal inequities refuse Latinx and Indigenous people, disabled and neurodiverse people, and LGBTQ people, among other groups, the right to curiosity.[15] We think deeply of all the underrepresented scientists and theorists whose innovative work is repeatedly passed over.[16] And we ask, How might we thatch our lines together? How might we help this forest grow? What we offer here is but one line of flight—one by which we stretch our brains and our hearts out beyond the brink of the world.[17]

1 THE SCIENCE OF CURIOSITY

He picked up his spade . . .
—Frances Hodgson Burnett, *Secret Garden* (1911)

You were curious when your hand reached out to pick up this book. What is that curiosity? Is it animal, vegetable, or mineral? Is it ethereal or tangible? Like a loris, can it be categorized and classified, or like love, is it difficult to define? It feels impossible to choose precise words to answer these questions. And yet curiosity seems as if it *should* be definable because . . . well, because . . . because it is so simple. And perhaps it seems so simple because it is so common.

Curiosity is everywhere around us—in the child sniffing the wavering wildflower, the youth playing a glossy new track, the adult pulling a dusty volume from the top shelf. It is so pervasive that a medieval mind might wonder if curiosity is simply in the air about us. Do we imbibe it, fill our lungs with it, and then share it with one another as we exhale? Does the air itself make us curious because its flittering molecules house the essence that *is* curiosity?

Perhaps. As English writer, philosopher, and clergyperson Joseph Glanvill suggests, air is certainly curious in the sense that it takes interest in its surroundings. Air's interest in the vibrations of pendulums leads to a bending of its airy body, and that bending serves to propagate sound.[1] The fruition of air's bending, the sounds themselves, appear curious in the nature of their movements, seeking to fill spaces and to mold volumes of air. Movements like these and many others reflect the nature of a being's curiosity. The fingers of a flutist appear curious as they explore the available keys to shape the most pleasing sounds. The feet of a walker appear curious as they explore the available terrain to circumnavigate the most awing vistas. The mind of a human appears curious as it explores the available knowledge to isolate the most engaging thoughts.

The apparent simplicity and ubiquity of curiosity call into question curiosity's value. Like dust, rust, and must, if curiosity is everywhere, why should we care about it? Why read (or write) a book about curiosity? For millennia, humans have chosen to place value *not* on that which is common but rather on that which is uncommon. We place little value on the leaves of deciduous trees such as red alder, paper birch, and box elder—so plentiful that they cover the earth in a rather audacious abundance; instead, we prize species that are endangered such as baobab and monkey puzzle trees. Similarly, we place little value on common rocks such as gneiss, limestone, and shale; instead, we prize mineral crystals of benitoite, black opal, and larimar—so scarce that each appears in only a single spot on the planet. When choosing virtues to extol, we do not consider dispositions common to many animals, including the mantis and amoeba; instead, we choose those that are uncommon and indeed often difficult for most humans to attain, including humility, sincerity, generosity, and authenticity.[2] But do humans only and ever value that which is rare? Or is it possible that curiosity need not be rare to be valued? Perhaps curiosity is valuable in its potentiality (as opposed to its actuality), in what it makes possible, in the doors that it opens, and the possibilities that it brings to life. Perhaps curiosity is valuable because it is a process that can sometimes *lead us* to that which is rare: a curio of truth that we then place in our mental cabinet. Deeply accurate knowledge of our world is extolled as both rare and valuable, and curiosity leads us toward that knowledge through the relevant byways of insights, perspectives, and hunches. Curiosity allows us to arrive at knowledge about our fellow humans, about our world, and about our past and future. It is fundamental to inquiry of all kinds, either practical or theoretical, naive or educated, and about all topics, from the sciences to the humanities, from emotions to subatomic particles.

The relation between curiosity and inquiry is unary in the sense that it is fundamental. To be curious is (at least in part) to use the mind to inquire. Yet the relation between curiosity and inquiry is also binary, ternary, quaternary, and more generally multiary in the sense that to understand curiosity, we must inquire into inquiry, and then we must inquire into the inquiry about inquiry. And the answers we seek lie somewhere in the infinite reflections of ourselves and our minds as we stand juxtaposed between two facing mirrors. Rather than attempting to unpack each reflection, here we will focus on how curiosity is relevant for inquiry. Later in this chapter, we will turn to the complementary reflection: how inquiry is relevant to the study of curiosity.

How is curiosity used in inquiry? As a focal point, we might take inquiry regarding the natural world and then question how curiosity is used in the sciences. Science is sometimes thought to comprise the directed seeking of a particular piece of information in well-crafted experiments. But in fact science is just as much the practice and result of undirected observations, without any particular experimental apparatus.[3] The seeking of useless knowledge is one of the venerated bastions of the scientific enterprise in modern history. Although one could justifiably hark back to Aristotle, in more recent times Francis Bacon is perhaps one of the quintessential proponents of simple undirected observation.[4] He opens his book *Novum Organum* (or *True Suggestions for the Interpretation of Nature*) with the claim, "Man, as the minister and interpreter of nature, does and understands as much as his observations on the order of nature, either with regard to things or the mind, permit him, and neither knows nor is capable of more."[5] A particularly compelling visual that lauds undirected observation is the frontispiece that Bacon selected for his *Instauratio Magna* (1620), where Flemish engraver Simon van de Passe etched two columns standing on a rocky shore, beyond which a ship sails into a vast expanse of water.[6] No ripple appears more valuable than another; no molecule is more truly water than another. Each observation, all observations are equivalently necessary in the labor of transforming the unknown into the known.

Such undirected observation takes an equal place in the *Philosophical Transactions of the Royal Society of London* (first published in 1662) alongside mathematical proofs and reports of laboratory experiments. The earliest such anecdotes report milk in human veins, the existence of "a very odd monstrous calf," and an account of a strange spring in Westphalia.[7] Some articles even provide predictions about what observations might be possible in the future, such as that titled "Monsieur Auzout's Speculations of the Changes, Likely to Be Discovered in the Earth and Moon, by Their Respective Inhabitants."[8] Ian Hacking, a Canadian philosopher, refers collectively to these items as "traveler's tales," emphasizing that anecdotes from undirected observations were a key component of the emerging genre of scientific journals at the time and an important component of Bacon's case for curiosity.[9] In contrast to these early priorities, the pressures of the ensuing centuries induced investigators to practice a utilitarian curiosity, chiefly by tackling questions of practical use. In the early 1900s, the strain was so great that educational leader Abraham Flexner founded the Institute for Advanced Study as a place where a scientist could once again think about questions that were simply interesting, irrespective of their utility. In his now-classic essay "The Usefulness of Useless Knowledge," Flexner suggests that our conception of what is useful may "have become too narrow to be adequate to the roaming and capricious possibilities of the human spirit."[10] At its best, science roams all unknown waters, with ethics at its helm and sails spread with the processive value of coming to know rather than the possessive value of owning the known.

CURIOSITY AND THE SCIENTIFIC METHOD

The curiosity to think, to collect, and to observe is germane to the process of scientific inquiry. In the same essay, Flexner writes, "Curiosity, which may or may not eventuate in something useful, is probably the outstanding characteristic of modern thinking."[11] In guiding the acquisition of both useless and useful knowledge, curiosity is a scientific tool to search for truth and to overcome uncertainty. But although necessary, curiosity may not prove sufficient to obtain the certainty that scientists often wish for. To appreciate this point, let us go back to Glanvill circa 1661. He had just published *The Vanity of Dogmatizing*, which pleaded for religious tolerance over persecution and freedom of thought over scholasticism. In the treatise, he questions how we can be dogmatic about religion (and the *immaterial* world) when we have such deep uncertainty about science (and the *material* world).[12] He builds his case by describing how incredibly ignorant humans are, demonstrating that we cannot explain our senses, discussing our incapacity to decipher human memory, revealing how little we know about matter, and underscoring our inability to understand even simple actions like the motion of a wheel. (Note that this latter bit of ignorance would be significantly ameliorated just a few years later, when Isaac Newton published his *Philosophiae Naturalis Principia Mathematica* in 1686.) With such little understanding, Glanvill asks, who are we to police one another's thoughts and beliefs?

This ignorance of ours comes in two kinds. It may be contextual or it may be intrinsic to human nature. Contextual ignorance can arise when a culture does not yet have the scientific tools needed to convincingly answer a scientific question, test a hypothesis, or build a justifiable theory. Intrinsic ignorance arises when an organism (e.g., human) is built with fallible machinery (e.g., perception or reasoning) that always keeps the truth at arm's length. Glanvill appears to be thinking of the second type: intrinsic ignorance. Our deep ignorance, he claims, is due in part to the fact that our affections mislead us; we love our own ideas and those of others, and we pay undue homage to antiquity and authority. These traits are human traits and impact our ability to see the truth, independent of the (im)perfection of our scientific tools. Geologist T. C. Chamberlain appears to agree. In 1897, he writes that we still fall prey to our affections.

> The moment one has offered an original explanation for a phenomenon which seems satisfactory, that moment affection for his intellectual child springs into existence, and as the explanation grows into a definite theory his parental affections cluster about his offspring and it grows more and more dear to him. . . . There springs up also unwittingly a pressing of the theory to make it fit the facts and a pressing of the facts to make them fit the theory.[13]

Scientists are humans, and science progresses in a manner that is at least partially hamstrung by human limitations.

As contemporary scientists, we increasingly appreciate the limits of our own knowledge and the tools we have to expand that knowledge. Scientific inquiry is the search for a kind of truth. As scientists, we are driven by an existential desire for certainty, in part because we are faced with so much uncertainty. We are uncertain not just because the unknown is so vast but also because our (human) senses are imperfect and our (human) perceptions are biased.[14] In some sense, scientific inquiry is thus an effort to become inhuman. Or perhaps nonhuman. Or perhaps superhuman. At times it feels ironic that the entertainment industry depicts scientists as those who might someday—in some fantastic, unreal world—create the superhuman (think *Iron Man*); the irony arises from the fact that even the most miniscule and mundane action in a scientific experiment— pipetting some liquid, counting a few cells, or imaging the unimaged—is an action seeking to attain unhumanity, to create certainty from the machinations of our uncertain beings. But how can and do we become not human in all of these small ways? Well, we often attempt to do so by marrying the freedom of curiosity with the constraints of the scientific method: the rules of inductive inference. We begin with a question, break the question down into objects and processes, develop a way to measure those objects and processes, devise a hypothesis and a null hypothesis, devise an experiment to refute one of the hypotheses, and then carry out the experiment so as to get a clean result.[15] The sequence allows us to form conclusions from data by reasoning from particular facts to general principles.

The process of inductive inference can certainly help us to supersede the uncertainties caused by our imperfect senses and biased perceptions. But there is a problem. The process is not foolproof. By the imperfections of our senses, we only ever see through a glass darkly, perceiving bits of reality overshadowed by errors. Because we view the results of every experiment through the same dark glass with the same imperfections, we can never fully reach our goal. Scientists seek to circumvent this rather grim limitation in two ways. First, we choose or construct an unbiased measurement apparatus such as a microscope, sextant, or telescope, and then we place that apparatus between us (the human observer) and the world we hope to understand. If all evidence in favor of a hypothesis comes through a measurement apparatus rather than through our human perceptual faculties, perhaps we can be somewhat spared from our biases. Then we will see the evidence for what it is: objective measurements that will or will not support our theories. Of course, if we are honest with ourselves, we will admit that this is a somewhat rosy perspective, and our hope is only somewhat realistic. Yes, a measurement apparatus ensures that the scientific enterprise is less tainted by our human limitations, but no apparatus is perfectly unbiased, and neither is our perception of the resultant measurements. Second, we use an iterative approach of hypothesis and observation. We create a potentially infinite loop by building a hypothesis from an observation, using that hypothesis to develop a new experiment to obtain a new observation, and finally using that new observation to refine the hypothesis. We can walk this loop over and over again in our goal to step closer to the nearest bit of knowledge. While certainty will always be just out of our reach, we can nevertheless make progress toward it using experimental equipment and iterative inductive inference.

We as scientists spend our lives channeling our curiosity in the service of certainty. Yet along the way, we increasingly embrace uncertainty and celebrate the paths toward understanding, without expecting its conclusive acquisition. Our community commonly describes the PhD as a process not whereby we create new knowledge but instead whereby we learn how little it is that we actually know. We love the process of scientific curiosity not because it results in perfection but instead because it can get us a bit closer. Our progress is reminiscent of Zeno's paradox of the tortoise and Achilles, and also of Aristotle's passage in the *Physics*, where he writes, "In a race, the quickest runner can never over-take the slowest, since the pursuer must first reach the point whence the pursued started, so that the slower must always hold a lead."[16] We embrace that imperfect scamper in part by valuing the passage (e.g., the vaccine that is 90 percent effective or the antibiotic that kills some bacteria but not all) and in part by valuing the serendipity (e.g., the discovery of an antibiotic, a law, a treatment in the most unexpected of places). Becoming a scientist is like learning the "gentle art of tramping," which as travel writer Stephen Graham observes, is not about the extent of the landscape traversed but rather about how we live along the way. We walk the curious paths of investigation, inference, and inquiry not as a "stunt" (where the end is reached) or "personal ornament" (where the beauty is perfected) but instead in order to "linger in the happy morning hours, to listen, to watch, to exist," finding quality "in moments, not in distance run."[17] To a scientist, curiosity then is both a means of moving closer to the reality of the world and a means of lingering along a passage, to somewhere or to nowhere, or to an anywhere that serendipitously becomes a somewhere. It is the edge that straddles the divide between the more-than-human and the simply human.

It is in fact these imperfections—of current knowledge, of modern solutions, or of human capacities—that mark the potential and the privilege of science. They ensure that the path continues, that the way is forked and meandering rather than straight and predetermined. As scientists, we seek proximal certainty ("Here is the path!") just as much as distal uncertainty ("Who can tell where it might lead?"). We look askance at those who claim that they have the last word because we know Nature will speak on. And this is where we revel. Our playground is a world of unanswered questions, of puzzles all about us, and entering into us like droplets of water in the air we breathe. Science enthralls because each step of certainty brings us to the threshold of uncertainty. We reach out and touch the sheer cliff face of Nature's secrets; it is unscalable, and we love her the better for it. Each clearing in the woods brings us to the next thicket of opacity. Each transparent stream sits upon a rock bed of ambiguity. Each sheer breeze leads inevitably to the dark clouds of vagueness. It is this juxtaposition of opposites that gives words to the Siren's song. In fact, science is very much a song, a verse, a poem. It celebrates a complicated whole, in which a few piercingly lucid moments are carefully sculpted and set in relief against a larger and longer-lived bedrock of impression that lies somewhere between the meanings of words, an undercurrent that lies somewhere beneath the data of experiments. In poetry, we hear in the background the swirling connotations and corollaries, sedimented lines of comeanings, and fossilized records of concept histories; in science, we hear similar material resonances, eerie, and far away, as we listen to Nature's lyrics. And to make matters more intriguing, the phrases change, as eons pass and evolution marches on. The living infinitude of science leaves no room for regret. One need never worry that the last piece of the puzzle will be found, instilling the sadness of completion, the permanent loss of future puzzling. Likewise, one need never fear that the final page of an author's corpus will be read, soaking the reader in the grief of the nevermore. No. Here is the joyous promise of the never-ending.

CURIOSITY AND THE PARTS OF A SCIENTIFIC QUESTION

Curiosity can lead us toward that which is rare (a curio of knowledge), but it is not guaranteed to do so. Scientific curiosity is a symbiotic line of relation between humanity and the world; as such, it is at once constrained by the laws of Nature and confined by the powers of human inquiry. As the banks of a river both define and are defined by a stream of water, so curiosity both defines and is defined by the process of inquiry. The give and take begins with the forming of a question. What is a question? In science, a question is a statement of inquiry regarding objects and their processes—or more simply, things and what they do. Forming a good question is not easy; it requires a careful choice of parts and a precise posing of the processes whereby those parts depend upon one another. The procedure for forming a good scientific question is akin to diagramming a sentence, as is sometimes taught in middle school language curricula. A set of rules allows the student to transform a river line (of thought) into a leafy tree structure (of thought bits), with articles waving in the wind and prepositions sitting snugly at the branch points like small children, dangling their legs.[18] "I think quietly in a tree." *I*—long vertical line—*think*—under leaf—*quietly*—down branch—*in*—straight line—*tree*—under leaf—*a*. The student draws these diagrams on crisp, clean, unlined sheets of paper, evincing the explicit and most proximal dependencies between the parts of speech: adjectives leaf from the noun stem, while prepositional phrases branch from the words they modify. Similarly, a good scientific question makes explicit the dependencies between objects and processes. In the diagram, the physical matter we seek to investigate is the subject of the sentence, the physical process is the predicate, the alternative measurements of the physical matter are the predicate nouns, the confound variables are the adjectives, and the moderator variables are the adverbs. It is only from the correct distillation of an idea (the sentence) into its component parts and their relations to one another (the diagram) that we will be assured that our scientific process (sentence construction) accurately tests a reasonable hypothesis (reflects our meaning).

When diagramming our curiosity, we begin as we often do in a sentence: with the subject or the physical matter of interest. What physical matter should we choose upon which to build a hypothesis and search for an explanation? In making that choice, our biggest challenge is to pick the relevant scale of matter. Should we address questions of human cognition at the scale of subatomic particles and Feynman diagrams, or at the scale of bodily tissues and behavioral motifs? The question of which scale will provide the most satisfactory, or the most "true," explanation for any particular observable phenomenon is one that has plagued scientists for centuries.[19] And unfortunately the proposed answers have in general been rather divisive. The most common answer in some fields of science is that of reductionism: the notion that the processes occurring at large scales can only be explained by processes occurring at smaller scales, and those processes in turn can only be explained by processes occurring at even smaller scales, and so on, ad infinitum. Reductionism is limited for many reasons, but one particularly compelling reason is well articulated by theoretical physicist Philip Anderson in his 1972 article published in *Science* titled "More Is Different":

> The main fallacy in this kind of thinking is that the reductionist hypothesis does not by any means imply a "constructionist" one: The ability to reduce everything to simple fundamental laws does not imply the ability to start from those laws and reconstruct the universe. In fact, the more the elementary particle physicists tell us about the nature of the fundamental laws, the less relevance they seem to have to the very real problems of the rest of science, much less to those of society.[20]

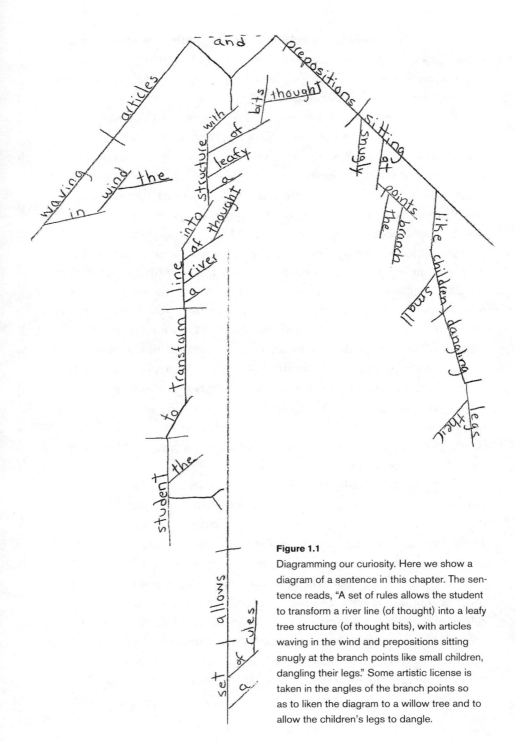

Figure 1.1

Diagramming our curiosity. Here we show a
diagram of a sentence in this chapter. The sentence reads, "A set of rules allows the student
to transform a river line (of thought) into a leafy
tree structure (of thought bits), with articles
waving in the wind and prepositions sitting
snugly at the branch points like small children,
dangling their legs." Some artistic license is
taken in the angles of the branch points so
as to liken the diagram to a willow tree and to
allow the children's legs to dangle.

Anderson joins philosopher Angela Potochnik and others in arguing that there exists no privileged scale for explanation. In fact each scale is characterized by its own laws, concepts, and generalizations, which cannot be explained by those of any other scale.

The notion that no single unit of analysis is relevant to all areas of science is admittedly freeing, but also challenging. How do we choose the unit of analysis with which to study any particular process? Well, we do so by considering three key properties that a unit of analysis in scientific inquiry could usefully have. First, ideally the unit would be something that we can reliably measure, or as reliably as possible given our version of Zeno's paradox described above. Second, the unit should be a concept in our hypothesis: the subject of our sentence. Together, these two features ensure that we begin our sentence in a reasonable manner. Importantly, once chosen, the unit of analysis determines the scale and nature of the other parts in our sentence: the physical process (verb), moderating variables (adverbs), confound variables (adjectives), and any alternative measurements of the concept of interest (predicate nouns). When formulated in language, the sentence is a hypothesis; when formulated in mathematics, the sentence is an equation. In both cases, the sentence represents a candidate model of the system. This latter observation leads us to a third key property that a unit of analysis in scientific inquiry could usefully have: namely, that it can be placed in a model that offers a parsimonious and accurate prediction of the object or process that it is meant to explain.

To appreciate the utility of these three criteria, let us consider the question of how to study, model, and ultimately explain a cat (take your pick: the Cheshire cat, Schrödinger's cat, or Granger's Crookshanks). Physicists are famous for modeling large mammals (e.g., cows) as simple geometric objects (e.g., spheres), and have no qualms using those same models for small mammals (e.g., cats). The spherical model of a mammal, large or small, is simple, tractable, and useful; a common question on a first-year college physics exam has something to do with a cow flying through the air. Yet such a model is also inherently simplistic. Why not add a few polygons? Triangles could be used to model the cat's ears, and a rectangle could be used to model the cat's tail. Of course, we could continue to add more polygons, tessellations, compartments, tissues, cell types, molecular machines, and eventually DNA, in a process moving from the coarse and simple to the fine and complex. The continuum of models illustrates an apparent trade-off between model accuracy and model parsimony. As physiologist Arturo Rosenblueth and mathematician Norbert Wiener once wrote, "The best material model of a cat is another, or preferably the same, cat."[21] Crookshanks is a model of the Cheshire cat, but a more accurate model of the Cheshire cat would be itself. Yet this more accurate model is far from parsimonious, and that lack of parsimony impacts the human capacity to understand and subsequently to know. Curiosity is an accurate model of curiosity, but have we learned anything from this model? Models that are both parsimonious and accurate can isolate the most important and most measurable components of an observable phenomenon, and they can describe the relations between components in a way that humans can understand. A model that is both parsimonious and accurate is one that has chosen a useful scale of curious inquiry, unit of analysis, and diagram of the question.

In this way, scientific inquiry and curiosity are profoundly interdigitated. Curiosity often leads us to the original question. Yet that question just as often begins as one that is rather ill posed and certainly naive. The constraint of what can actually be measured serves to hone the unit of analysis, and the strictures of formulating a hypothesis and its associated model serve to add precision to the question. In turn, curiosity presses back on the strictures, asking, What sorts of models might be true? What if my noun actually did y instead of x? Could the equation be written this way rather than that way, and if so, what would that formulation mean for my prediction? Walking the edge between humanity and nature, scientific curiosity guides the river of inquiry while also pushing against the banks to expect the unexpected.

THE SCIENTIFIC STUDY OF CURIOSITY

Now that we have unpacked how curiosity is relevant for science, we turn to the question of how science is relevant for curiosity. Suppose we are curious about curiosity. What is curiosity? What are its component parts and processes? How do we choose a unit of analysis or scale of inquiry? What measurement apparatus should we use? How could we formulate the model? How could we construct and refine an initial hypothesis? What sentence do we wish to write and how do we diagram it? In posing and seeking answers to these questions, scientists are often driven by distinct goals. Some scientists are most compelled by utilitarian values. Prior work has linked curiosity to happiness and innovation—two phenomena that are highly desirable in many cultures and societies. Curiosity is correlated with flourishing and physical activity, and is anticorrelated with depression.[22] Moreover, curious people are often innovative people who have the capacity to be world changers by imagining a different future.[23] Were we to understand what curiosity is, might we better support or even enhance it? Other scientists are most compelled by their own desire to know, fully independent from any utility, which Flexner so eloquently described as the "expression of the untrammeled human spirit."[24] Perhaps we wish to understand curiosity simply because it is curious. It is strange, puzzling, and mystifying. We cock our head in a quizzical stare. What is this oddity before us?

The modern scientific study of curiosity began in earnest in the field of psychology. In this discipline, practitioners notice the physical and mental behaviors of human and nonhuman organisms, and seek to understand them. One particular behavior that is often thought to epitomize curiosity is the collection of items in a so-called cabinet of curiosities. Consider Ole Worm, or the Latinized form of his name that he was known to use on occasion, Olaus Wormius. A natural philosopher in Copenhagen in the early 1600s, Worm collected rare minerals, plants, animals, and artifacts in a space he called the Museum Wormianum, and he wrote a four-book catalog called the *Museum Wormianum* for its contents, a compendium of descriptions and speculations. What induced Ole to collect, to catalog, and to curate?

Arguably, curiosity. But to press deeper, perhaps we could ask the question, What spirit or entity drove him to create the Museum Wormianum?[25] Let us begin by assuming that the cause lies within the body—perhaps in his heart, his feet, his hands, or his mind. True, his heart allowed him to live but did not specifically drive his curiosity. His feet are more promising:

they walked his body to any given artifact. His hands picked up the artifact from the ground or ledge or another's hand. His mind desired to know the meaning or use of the artifact. But which is more proximal to curiosity? In the chain of temporal precedence, the desire to know predated the walking feet, which in turn predated the grasping hand. So in some sense, the desire to know is the first cause in this chain of events. The mind appears to be the source.

What happens in the mind when we are curious? What is curiosity in the mind? How can we unpack the mind's curious elements? With philosopher Gottfried Wilhelm Leibniz, in his "Towards a Universal Characteristic" (1677), we seek an alphabet of human thoughts: "There must be invented . . . a kind of alphabet of human thoughts, and through the connection of its letters and the analysis of words which are composed out of them, everything else can be discovered and judged."[26] In the mid-1900s, Daniel Berlyne, a British psychologist, sought to distill such an alphabet—not for all of thought, but for curious thought. First, he noted that the alphabet did not contain a letter for informational utility. He wrote, "Many of the queries that inspire the most persistent searches for answers and the greatest distress when answers are not forthcoming are of no manifest practical value or urgency."[27] Humans were curious whether or not the information they sought was useful to them. If utility was not driving curiosity, then what was? Berlyne offered perceptual curiosity as an alternative and defined it as a drive that is aroused by novel stimuli (Look! A quaflet!) and reduced by continued exposure to these stimuli (Oh no, not another quaflet? ::yawn::).

Berlyne further contrasted perceptual curiosity with epistemic curiosity, which he defined as a drive reducible by the reception and subsequent rehearsal of knowledge (What? A quaflet is a spindly legged, gnat-sized insect shaped like a Viking drinking horn, most well-known for releasing a liquid drop of existential realization—of expansive self-revelation—into the inner corner of the left eye of unsuspecting daydreamers? ::turning to friends:: Becket, Loren, did you know that a quaflet is a . . .). Whereas perceptual curiosity might be characteristic of many nonhuman animals, epistemic curiosity was thought to be largely reserved for humans. Berlyne further distinguished specific from diversive curiosity; the first referring to a desire for a particular piece of information (What do quaflets release into the outer corner of the right eye of suspecting daydreamers?) and the latter referring to a desire for any information (Tell me about quaflets, do!). The distinction between specific and diversive curiosity is reminiscent of Virginia Woolf's account of the person who loves learning versus the person who loves reading: the former "searches through books to discover some particular grain of truth upon which he has set his heart," whereas the latter simply manifests the "humane passion for pure and disinterested reading."[28] The two axes of epistemic to perceptual and specific to diversive together form a two-dimensional plane, with each distinct quadrant describing a wide range of exploratory behaviors in humans.

On the backdrop of distinct drives, Berlyne sketched out the sequence of events in the thought-line of curiosity: the letters of the alphabet that together form the word or the parts of speech that together diagram the sentence. Following US psychologist and behaviorist B. F. Skinner, he reasoned that the sequence should begin with a *thematic probe*, which is any stimulus that elicits not a single thought but rather a train of them. The thematic probe itself can be broken down into two parts: a cue stimulus, which supplies the starting point, and a motivational stimulus, which limits the subsequent thoughts to the relevant portion of knowledge space and simultaneously motivates the human's response. Questions are quintessential thematic probes, containing both cue and motivational stimuli. Consider the question, "Why is a clump of dune grass—on which a quaflet rests—surrounded by an indentation in the sand at a fixed radius from the grass's center, as if a finger had drawn a circle around it?" The cue stimulus is a clump of dune grass surrounded by an indentation in the sand. The motivational stimulus is the phrase "why is," which limits the train of thought to why concepts and evokes a motivation to answer the question. The fact that the question limits the ensuing thought space is consistent with the notion that there exist conditions that define the useful versus not useful answers to why questions.[29] Or consider the question, "How does wind draw a circle in the sand?" The cue stimulus is wind drawing a circle in the sand and the motivational stimulus is the phrase "how does," which limits the train of thought to how concepts, and drives the human to infer that the wind has bent the dune grasses and swept them around such that they act like fingers drawing a circle in the sand.[30]

Why is a motivational stimulus motivating? Why does a thematic probe induce curious thought? Why do we follow the thought forward? Is there something rewarding waiting at the finish line that drives us tramping onward? Is an end reached or a beauty perfected? In considering the idiosyncrasies of curious thought, it seems that even if there is a treasure waiting at the end of the rainbow, it does not appear to be something like a pot of gold that everyone would value. While the quaflet drives Lyle, the dune grass motivates Sophie, and neither ignites a spark for Ira. In a seminal theory paper in the 1990s, US educator and economist George Loewenstein noted that the information people seek appears in fact to have no particular value: "The theoretical puzzle posed by curiosity is why people are so strongly attracted to information that, by the definition of curiosity, confers no extrinsic benefit."[31] But if there is no benefit, why be curious? To solve this puzzle, Loewenstein developed what he called the *information gap theory*. This theory offers an explanation of idiosyncratic curiosity that is both specific (seeking a particular piece of information) and epistemic (seeking information that is of knowledge rather than of perception). Loewenstein claims that the information gap exists between what one knows and what one wants to know. Then the intensity of curiosity for a piece of information tracks its ability to close the information gap, thereby resolving uncertainty. The end we reach is what we have been seeking all along: certainty. The beauty perfected is the crystallization of knowledge. Thus the information obtained is valuable in its potentiality (as opposed to its actuality), in what it makes possible; it opens the door to understanding. What is valuable is that the puzzle piece fits into the gaping empty spot. To fill that function, the puzzle piece need not be made of gold or benitoite, black opal or larimar. It must simply have the right shape to draw certainty from uncertainty.

With theory in hand from Berlyne, Loewenstein, and many others, the inquisitive scientist is now ready to place a measurement apparatus between themselves and the world, and to use the resultant measurements to evaluate the theory. One particularly common measurement apparatus used in the psychology laboratory is called a scale.[32] A scale is a set of questions developed to measure a well-defined construct or idea. Before use, the scale is typically evaluated for validity and reliability, and iteratively refined until the validity and reliability reach certain levels accepted by the scientific community.[33] For example, one can choose a definition for curiosity and then develop a set of questions to measure that kind of curiosity:

For each of the following statements, choose *strongly agree*, *agree*, *neutral*, *disagree*, or *strongly disagree*.

1. I enjoy learning about subjects that are unfamiliar.
2. When I come upon an incomplete puzzle, I imagine the final solution.
3. I think about many different answers to the same question.

Valid questions are those that probe curiosity rather than some other construct (e.g., bibliophilia, bubbliness, or bombastry) and reliable questions are those that a person will answer the same way each day, thus accessing a trait of their personality.[34] After a volunteer marks their responses to each question, the responses are coded and summarized into a score representing the volunteer's personal level of curiosity, whether high, low, or middling.[35] The scale can also be used to assess evidence in favor of the theory of curiosity from which the definition arose, including evidence for the theory's accuracy and parsimony. To evaluate Berlyne's theory separating curiosity along two main dimensions (epistemic-perceptual and specific-diversive), for example, one could begin by creating one scale for each dimension and then one could assess whether those two scales better explained the range of real-world curious behavior observed in humans than a competing set of dimensions. To evaluate Loewenstein's information gap theory, one could develop a scale to measure the gap that exists between what one knows and what one wants to know, and evaluate the relation between that gap and the degree of curiosity a person manifests to fill the gap.

This is the psychological science of curiosity: it requires the crystallization of a theory, a concretization of that theory into measurable units and processes, and the development of instruments (such as scales) to measure those units and processes to provide evidence in favor of—or that contradicts—the theory.

MIND OR BRAIN?

Let us return to Ole Worm and his Museum Wormianum in Copenhagen in the 1600s. Earlier we had asked about the first cause of his curious actions. We shelved his feet and hands, and landed on his mind, which motivated our foray into psychological science. Resurrecting the question, we can now ask whether we are satisfied with our initial answer. It is arguably true that Ole Worm's desire to know led him to walk toward the artifact and subsequently led him to pick up the artifact to bring it back to his cabinet of curiosities. It is also true that the desire to know is said to arise in the mind. But can the mind be the *cause* of the desire to know?

To answer this question, let's think a bit harder about what cause means. Several distinct types of cause can be fruitfully defined, but a particularly useful one for our investigation here is the following. Let us consider a cause that is defined in temporal and spatial relation to an effect.[36] Here, cause and effect are related temporally because an effect cannot occur before its cause, and a cause cannot occur after its effect. Similarly, cause and effect are related spatially because the cause is located somewhere in space, and it works through space to create an effect. The quaflet is the cause of the daydreamer's existential realization because its porter-colored liquid first drops from its body, then moves through the air, and finally lands in the inner corner of the left eye to do what chemicals do. Defining a cause by its temporal and spatial contiguity to the effect is intuitive. When we purposefully cause something in our world (or perceive something to be caused), we observe that the effect occurs after the cause, and we observe that the relation between cause and effect is mediated by space. Moreover, our experience tells us that an event happening at some much earlier time and in a much different space cannot reliably be proven to have produced the effect we observe here and now; it is possible that many other causes and effects have intervened in the meantime, and in the meanspace. With this notion of cause in hand, we can return to the relation between mind and cause.

Does the mind have a physical location, and can it use physical processes, embedded in time and space, to cause the actions and movements of curiosity? Does the desire to know have a physical location? Does a single thought have a physical location? These questions drive us to the physical substrate of the mind: the brain. As an organ, the brain is a tissue, within which individual neural events have a precise physical location. The brain employs electrical signals, which are physical processes embedded in time and space, to drive a wide range of effects such as those behaviors that we as humans commonly attribute to curiosity. If we are seeking the first cause of curiosity, might it behoove us to consider the brain? What might a brain cause of Ole Worm's museum be? Perhaps there is a curiosity area in the brain, and it sends an electrical signal along a bundle of neuronal axons to the motor area of the brain. The motor area in turn sends an electrical signal through the peripheral nervous system to the feet, inducing them to walk toward the subject of wonderment. The ensuing eye-foot coordination is a bit of a ping-pong between the peripheral nervous system and the central nervous system, and between the visual cortex (feeding the eye) and the motor cortex (eyeing the feet). A similar motoric signal guides the hand to grasp, and the subsequent visual, auditory, tactile, and epistemic information gained from the subject satiates the curious impulse, filling the gap in knowledge. Is this how it all works?

THE NEUROSCIENCE OF CURIOSITY

Contextual ignorance held us back from answering many of these questions until the latter half of the 1900s, around the time when the first neuroscience departments in the world were being founded. A surprisingly young discipline, neuroscience is the science of the nervous system. Scholars in this discipline seek to identify and understand the neural processes subserving being and behavior. In the context of curiosity, they aim to identify the areas, interareal connections, and informational processes that explain the behavior of seeking information.[37] Common instruments measure (1) anatomy, in the form of morphology (or shape), tissue characteristics, cytoarchitecture (or the arrangement of cells in a tissue), and the distribution of physical tracts linking different bits of cortex, and (2) physiology, in the form of oxygenated blood flow, electromagnetic fields, and electrical currents. Those instruments, like the scales in psychology, can be used to mitigate the limitations of our biased human senses and to evaluate the evidence in favor of a hypothesis, using the methods of inductive inference. Let's envision such a process.

What evidence could neuroscience offer for the claim that the brain is a useful "first cause" of curiosity? Ideally, to gather precise evidence for a complete causal chain, we would fully and simultaneously measure neural activity in the brain (the central nervous system) and neural activity in the body (the peripheral nervous system). That simultaneous and complete measurement would allow us to watch—with spatial and temporal contiguity—how the activity in one part of the brain might lead to activity in another part of the brain and down through the spinal cord to move the body. This kind of full measurement is not possible with our current neuroimaging technologies, nor is the measurement of even a single brain region if the curious individual is ambulatory. In fact, even if a person is motionless, we still have no way to *directly* measure the activity of a single brain region; typical studies make *indirect* measurements of oxygenated blood flow as a person is lying down inside a magnetic resonance imaging (MRI) scanner. Without access to a full casual chain, we might instead seek a correlative chain (A is correlated with B, which is correlated with C); correlative chains can be consistent with (but not proof of) a causal chain. In the first step of the chain, perhaps we find that activity (A) in a specific area of the brain (our "curiosity area") is correlated with a cognitive state (B), such as the desire to know. In the second step of the chain, perhaps we find that a cognitive state (B) is in turn correlated with actions to seek information (C), such as walking toward an artifact and stooping to pick it up. Then we can test whether the neural activity (A) is correlated with actions to seek information (C). The *correlative* chain from neural activity to a cognitive state to an action can provide evidence in favor of a *causal* chain of curiosity in which the brain is the first cause.

How do we choose an appropriate measurement of neural activity? We can diagram the sentence by identifying and separating out the unit of analysis (subject), associated processes (verb), and other relevant variables (prepositions, adjectives, and adverbs). By diagramming the sentence, our goal is to formulate a hypothesis using language, and to devise an equation using mathematics, in such a way as to ensure that the units are measurable, the model is accurate, and the model is parsimonious. Importantly, assessing the model's accuracy can guide us toward a useful choice for the neural measure. If of three candidate curiosity areas X, Y, and Z, only one's activity is strongly correlated with actions to seek information, then we choose that one and ignore the others. Similarly, assessing the model's parsimony can guide us toward a useful choice for the neural measure. A model made from the activity of three areas, in some gourmet mathematical soup of integrals and derivatives, is less parsimonious than a model built from the activity of one area, in a simple sandwich of subtraction; if both models are equally accurate, then we choose the one-area model.

As we move through this process, our curiosity about curiosity calls out an initial inquiry, and then Nature answers our call with guiding words as we see which models fit well and why. Our relationship with Nature feels a bit like the relationship between a reader and an author. We attempt to read the book that Nature has written. In fulfilling our role in the relationship, we heed Woolf's admonition, "Do not dictate to your author; try to become him. Be his fellow worker and accomplice."[38] After all, how could we dictate to Nature when we often do not even know her language? But in being Nature's accomplice, our vocabulary grows, and our understanding of grammar and syntax deepens. And the cadence of our sentences—as they transform from vague river to precise tree—inch closer and closer to those that Nature has written; we reshape, revise, and reform our models by working in communion with (not conquest over) the natural world.

Using experiments of this flavor, neuroscientists have identified two neural circuits that support curious behavior: the reward circuit and the cognitive control circuit.[39] The reward circuit and associated motivation areas span the cortex, which is largely located in the brain's exterior, and the subcortex, which is largely located in the brain's interior.[40] This circuit is active during curious search and sampling, and partially reflects the motivation to obtain novel information, the expectation of reward, and the feeling of reward when that novel information is obtained.[41] In contrast, the cognitive control circuit supports goal-directed cognition, and contains brain areas largely located in the cortex that control the timing and nature of both search and sampling.[42] Evidence suggests that undirected exploration is particularly well supported by the medial frontopolar cortex, which monitors whether cognitive resources should be maintained toward the current goal or redirected toward others.[43] The lateral frontopolar cortex is implicated in directed exploration, monitoring several goals for maintenance or redirection. The posterior prefrontal cortex is associated with the exploitation of acquired knowledge, implementing cognitive control to optimize the performance of the current goal. Together the reward and cognitive control circuits enact processes associated with seeking and acquiring information.

In the pursuit of the neural underpinnings of curiosity, two key challenges remain. First, there remains a gap between locating the neural activity associated with specific behaviors and understanding the how or why of that association. Obtaining the latter is difficult. Second, these same circuits appear to support many other (noncurious) behaviors observed in humans. Hence work is needed to fine-tune the model, iteratively assessing model accuracy and parsimony.[44] These challenges ensure that the neuroscience of curiosity is a subfield of inquiry that will continue to motivate scholars well into the future.

*　　*　　*

As we close this jaunt into the curiosity of science and the science of curiosity, it is worth pointing out several limitations and open questions. While complementary efforts exist in neuroethology, behavioral science, and nonhuman animal research, here we have focused on the psychology and neuroscience of *human* curiosity, with an emphasis on psychometric scales and fMRI studies. Against that backdrop, a key difficulty in the psychological science of curiosity is the fact that the self-reports solicited by scales can be clouded by the person's perceptions of the traits that they think are valuable in their society and culture. As Loewenstein put it, "Asking people how curious they are (and most of the other items included in curiosity scales) makes the purpose of the scale obvious to subjects. This is a serious deficiency when one is measuring a trait that is widely recognized as socially desirable."[45] Humans can respond to a scale in a way that reflects their perception that curiosity is currently a socially desirable characteristic rather than in a way that is truthful to their own selves. One could argue that neuroscience is more objective in that it does not rely on self-reports and thus should be less influenced by the person's perceptions. Neuroscience also has the benefit of having physically located units, such as neurons and brain areas, and a physical network of informational tunnels and drains, such as synapses between neurons or white matter tracts between brain areas.[46] And of course, the physicality of the brain is consistent with common scientific notions of cause and effect being temporally and spatially contiguous. But the advantages of neuroscience over psychology are balanced by its disadvantages. While we can measure neural activity, are we any closer to thought? While we can track the flow of electrical current, are we any closer to the desire to know? While we hold the brain in our hands, the mind hovers above us chuckling.

The discipline-specific limitations of psychology and neuroscience aside, there are some quite generic existential concerns voiced by both recent and current practitioners. For example, Loewenstein despairs of ever being able to define the cause of curiosity: "The remaining question—the cause of curiosity—is inherently unanswerable."[47] In a similar ilk, scholars of the brain somewhat radically suggest that "one factor impeding our understanding has been too much focus on delineating what is and is not curiosity."[48] In other words, we should give up (for now) on attempting to define the phenomenon of interest. Perhaps instead we are still in the exploratory stages of reporting an odd spring here and a monstrous calf there, and are not yet ready for the practice of diagramming a sentence. "We recommend that the definition stage should follow a relatively solid characterization of curiosity, defined as broadly as possible."[49] Should we constrain ourselves to descriptive characterizations and not seek a guiding definition? But perhaps the problem is not that the cause of curiosity is unknowable nor that the definition of curiosity is unattainable but rather that the disciplines currently used to develop a definition are not up to the challenge. What if we were to turn from the natural sciences to their precursor, philosophy? Could a more circumspect and historically grounded assessment of the workings of the mind provide a novel definition of curiosity that could direct the future of scientific inquiry into its nature?[50]

2 CURIOSITY AS EDGEWORK

We strangely stand on,—souls do,—on the very edges of their own spheres, leaning tiptoe toward and into the adjoining sphere.
—Ralph Waldo Emerson, *Journals and Miscellaneous Notebooks,*
1843–1847

What is curiosity if not standing on the edge and peering out beyond it? Or leaning tiptoe toward something just beyond one's reach? Dissatisfied with the boundary, the border, the bubble in which one finds oneself. Indeed, what is curiosity if not the edging of spheres toward one another, so as to bridge and to blur and to build? Curiosity at its liveliest is hardly satisfied with sussing things out only to bring them home to roost—lining the crevices of one's mind with settled things and letting them take up space alongside other things one need never reconsider.

No, curiosity has a penchant for striking out yonder. Pitched ever so slightly forward, it repeatedly casts a line as if stretching and inching its way around the world—a world that is itself a "teeming edge-world."[1] In this sense, curiosity cannot simply be the need to know (which would then be satisfied with knowledge gained) or the motivation to seek information (which would then be satisfied with information secured). Rather, curiosity has to be an extending and extensive instinct, always edging around. And indeed, what could be stranger?

Or at least that is what we propose, and we have good reason for doing so. For while curiosity has largely been understood on the model of epistemic acquisition for millennia, other conceptual possibilities are buried in that same history. Here we turn, then, from the neuroscience and psychology of curiosity to the philosophy of curiosity. We do so in order to trace the roots of the acquisitional narrative back to its beginnings, but also to tell a different tale, a story of curiosity in relation, of curiosity on the edge.

Long before psychology used qualitative and quantitative methods to investigate the science of curiosity as a mental state or a personality trait with corresponding behaviors, there was philosophy. And long before neuroscience used neuroimaging and data science to understand the neural circuitry and systems that precondition curiosity, there was philosophy. Indeed, the framework of information seeking that so often defines psychological and neuroscientific studies of curiosity finds its roots here. Philosophy is, in some sense, the Ur-discipline, the source point of reflective engagement with the world. As such, it frequently carries the kernels of contemporary thought. And yet across its long and messy history, philosophy also houses all kinds of other possibilities for thinking differently— possibilities that surfaced here and there, but never got codified into paradigms or sedimented into theories; possibilities that still lie dormant.

In the academy today, philosophy is a discipline that uses conceptual analysis and logical argument to suss out the nature of things, the nature of knowledge, and the nature of value. Philosophers of curiosity might naturally ask then, What is curiosity? What is its role in the construction of knowledge? And under what conditions does curiosity contribute to or detract from a good life? Inquiries like these are in fact underway.[2] More than analysis and argumentation, however, philosophy is also a discipline of concept creation. It is, at its core, an attempt to think differently. Beyond plumbing curiosity within current ontological, epistemological, or ethical frameworks, philosophy pushes one to ask how curiosity has been conceptualized over time and how it might be conceptualized otherwise. What are the limits of the term as it has been traditionally conceived? What are its hidden investments, so often obscured through commonsense usage or ideology? What about it deserves greater attention and closer analysis? Beyond pursuing certain methods and topics of inquiry, then, philosophy critically assesses intellectual lineages and explores new possibilities for thought. This work involves deconstructing existing concepts and generating altogether fresh ones, not safeguarding universals, but submitting ideas to the endless process of renewal, replacement, and mutation. As Gilles Deleuze and Félix Guattari write, "Philosophy is the art of forming, inventing, and fabricating concepts."[3] That invention, through and beyond historical resources, is the project of this chapter.

In what follows, we revisit the long philosophical history of conceptualizing curiosity as an individual desire to know or, in contemporary scientific nomenclature, a drive for information or to fill knowledge gaps. We take a moment to appreciate what this traditional conceptualization has allowed to surface, but also what it has hidden. Turning to take an expressly relational approach, we then propose a new conceptualization of curiosity, according to which the curious impetus is not knower Y's desire to know X but rather a practice of connecting, of building relations between and among systems of (un)knowers, (un)knowledges, and things (un)known.[4] We then dive back into the history of philosophy to excavate inklings of this novel concept in past theoretical constructions before exploring the implications of it for psychology and neuroscience. Ultimately, we explore the potential payoffs of this dramatic paradigm shift. What if, in network nomenclature, we understood curiosity less as an individual capacity to build or acquire nodes, and more as simply the energy of the edge? What would this allow to be seen and noticed, tested and perturbed?

A few notes on method before we begin. First, in this chapter we consciously undertake a *genealogical* investigation of the philosophical sort. That is, we analyze how what is currently thought has its roots in prior thought, but also how thoughts of yesteryear hold underappreciated resources for thinking things differently today. As such, in the genealogical investigation to follow, we unearth resources in the past for new present and future conceptualizations of curiosity. Second, in that work of diving back into the past, we track the word *curiosity* in several languages of Western philosophy (e.g., Greek, Latin, French, German, and English), but also alongside the words "connection," "link," and "relation." Where these terms appear together, a different paradigm lies in wait. Third and finally, we understand a network to be any structure with netlike intersections and interstices. As modeled by graph theory, a network is composed of nodes (loci) and edges (relations between the loci). Allowing contemporary network sensibilities to resonate with past relational paradigms, we therefore attend to conceptualizations of curiosity as inherently interconnective rather than individualizing—conceptualizations of curiosity not as nodal acquisition but as edgework.

CURIOSITY AS NODAL ACQUISITION

For millennia, curiosity has been characterized predominantly as an individual impetus to secure pieces of information. Whether it is Saint Augustine's "lust to experience," John Locke's "appetite after knowledge," Sigmund Freud's "scopophilic drive," or Hans-Georg Voss's "motivation to explore," philosophers and psychologists of the Western intellectual tradition have largely understood curiosity at the level of the organism, its forces, and its feelings. Within this framework, curiosity is the desire, appetite, impulse, interest, motivation, feeling, or drive state to know X, understand X, find out X, better cognize X, fill an information gap with X, or gain information about X. It is individual and it is acquisitional, rooted in units of knowers and units of knowledge. Such characterizations are informative and useful for a variety of philosophical and scientific investigations into the nature of human knowledge, but they also have distinct limitations. In this section, we explore the possibilities and parameters of this tradition.

In the ancient and medieval periods, curiosity was held in deep suspicion as a largely vacuous interest in novel or forbidden information. The curious person was thought to reach for what was hidden and secret, beyond their proper purview, and thereby isolate themselves from spiritual and social community. The accounts of Plutarch, a Roman essayist, and Augustine, a Christian bishop, are representative in this respect. For Plutarch, the curious person—or in his nomenclature, the busybody—was an oddball, a singular fellow and a bit of an earwig. The busybody's curiosity manifests as "a desire to learn" and "a passion to find out" what is truly new and novel. They push into the bazaars, marketplaces, and harbors asking, "Is there any news?" But because middling news is easy to come by—insofar as people shout it from the rooftops or splatter it across their social media feeds—real news is hard to get. What remains truly novel is typically "hidden," "obscure," and "concealed" because it is troubling: the dirty secrets and the lies. They are hardly timid about sussing these out. As Plutarch writes, the busybody "strips off not only the mantles and tunics of those near him, but also their very walls; he flings the doors wide open and makes his way, like a piercing wind."[5] The busybody's quest for the maggoty mysteries is pursued without concern for themself or others. Abandoning their own interests and leaving their own soul in seedy squalor, the busybody jeopardizes their friendships and destroys social ties. As an antidote to this "disease" of curiosity, then, Plutarch recommends a suite of ascetic practices, including not opening a letter upon its receipt, not consummating a marriage, and, upon hearing a theatrical performance in the distance, walking in the other direction. The message is simple: abandon the singular secret, the devil in the details.

For Augustine, who largely Christianizes the ancient account, the curious person is a sinner, the weaker brother, a sheep in need of a shepherd. Genesis begins, of course, with a story of creation and curiosity. The first woman, Eve, wants to partake of the one tree forbidden to her, representing the knowledge from which she and Adam are prohibited. Curious to know good and evil, she eats—that is, she reaches for, grasps, tugs, acquires, ingests, and consumes—a single piece of fruit from the forbidden tree and her eyes are opened. Through her, all of humanity is condemned, metaphorically, to creep on its belly like a serpent, eating things worldly rather than spiritual. "For one who eats the earth," Augustine writes, "penetrates things deep and dark but nonetheless temporal and earthly."[6] This now-generalized human curiosity manifests itself as "a lust for experimenting and knowing" a string of insignificant things of no real spiritual worth. Fallen humanity tumbles headfirst into a "buzz of distraction."[7] People frequent the theater, gawk at mangled corpses, obsess over dreams, signs, and wonders, titillate themselves with astrology and the dark arts, marvel at the stars, and court ghosts. They hunt not only creatures but also idle tales and scientific details for sport. Their eyes glaze over, glued to the minor drama of a spider or a lizard catching flies. In this burning need for bits of knowledge, isolated from spiritual meaning in religious communion and divine wisdom, each person is condemned to the very vacuity and exile of the knowledge they seek.

While the hyperindividualized nature of curiosity and its interest in orphaned details is of the utmost concern in the ancient and medieval periods, these same characteristics are curiosity's strongest commendations in the modern period. This is the era of early modern science and the individual, which then sets the stage for the liberal subject and the Industrial Revolution. Rather than novel or forbidden information, modern curiosity seeks eminently useful information. Philosophers René Descartes, Locke, and David Hume all offer similar accounts in this regard. Descartes defines curiosity as a "desire to understand."[8] Following upon a wondrous surprise of the soul at anything new or strange, curiosity moves to understand what is useful for the flourishing of the human organism and society. Likewise, Locke defines curiosity as "an appetite after knowledge." As the fundamental tool to remove our ignorance—or fill information gaps—curiosity naturally moves from asking "What is it?" to "What is it for?"[9] Lastly, Hume defines curiosity as "the love of truth." Echoing Descartes, he remarks that curiosity, once generated by "the suddenness and strangeness of an appearance," then focuses on what is "important and useful to the world."[10] In the spirit of Francis Bacon, then, modern curiosity, performed by independent knowers, courts surprising information for its own sake, a process that itself ultimately serves the cause of social utility. Singularly curious objects inspire the individual practitioner of curiosity to gain knowledge of eventual use to society.

The conceptualization of curiosity as an individual drive to find out an individual thing continues on into the present, as indicated by contemporary work in epistemology, moral psychology, and philosophy of language. Accounts by philosophers Ilhan Inan and Lani Watson, for example, characterize curiosity as an individual motivation to find out a specific referent or to acquire certain epistemic goods, respectively. Inan defines curiosity as an intentional state of mind expressible by an inostensible term.[11] Put more simply, for him, curiosity involves focusing on something specific that one can point to with one's finger, one's eye, one's tongue, or the sounds that roll off of it; and something that is nevertheless beyond one's knowledge, something one cannot yet fully describe or grasp. For instance, "when [Sherlock] Holmes is curious about who the murderer is," Inan writes, "his primary goal is to find the referent of the singular representation the murderer."[12] When the referent of the inostensible term is finally found, curiosity is satisfied. Similarly, for Watson, curiosity is a "motivation to acquire worthwhile epistemic goods." Such motivation arises from an awareness of one's real or presumed ignorance, and must be strong enough to proceed "at some marginal cost," involving some "small sacrifice." The epistemic goods to be acquired must be worthwhile—that is, nontrivial—sorts of "information, knowledge, and understanding." On Watson's account therefore, the "butterfly preservationist" is not curious whereas the "butterfly collector" is because the former is not motivated to acquire whereas the latter is. Furthermore, the person obsessed with celebrity gossip or content to "count blades of grass" is not virtuously curious because both interests are essentially trivial.[13] While Inan focuses on the linguistic conditions and Watson the virtue implications, both take curiosity to be an individual impetus to acquire definable pieces of information.

Across the history of philosophy, the traditional view has been that curiosity is subject Y's desire to know information bit X. Curiosity's personal ad, then, might read: Singular unitary knower seeks specific unit of knowledge. The difference across philosophical accounts and eras appears less in the who and what than in the how and why. Theorists diverge significantly in their description of the method and value of the curious subject's information acquisition. Whether the information is new, forbidden, useful, linguistically describable, or worthwhile, however, and whether it is sought via desire, appetite, ardor, interest, intention, motivation, or drive, curiosity is a subject's seeking of information. This traditional view thus illuminates the sorts of information one is curious about, the ways one is curious about that information, and the states in which one finds oneself when one is curious about that information. It is nonetheless limited in its precision and explanatory power by its individualistic and acquisitional commitments. How might communal and relational commitments change that view? What if the paradigm for curiosity were the butterfly conservationists who are in fact deeply involved in cultivating—and thereby

understanding—the ecological conditions that sustain butterfly populations? What is a view of curiosity that could account for the networks of
relations in which knowers, methods of knowing, and knowledges sit? And
what, indeed, is the energy of the edge?

CURIOSITY AS EDGEWORK

We are all familiar with the experience of acquiring knowledge—that is, of
wanting or needing to know something and, after some interval of time and
some investment of energy, getting that knowledge. It is an effective experience, or an experience of effectiveness, that is not without its pleasures.
But we are also familiar with the experience of incubating an inquiry—
that is, letting it grow, sinking its roots deep into our souls (for lack of a
better word), and stretching its branches into everything we think and say
for weeks, months, or even years (dare we say decades?). This is the kind
of inquiry through which we come to fall in love with a knowledgescape.
We visit it, sleep under its stars, and take up residence in its alcoves until
it becomes part of us, and we become part of it. This is no scurry across
the fence to poach a few pears. This is no snatch and skedaddle. This is the
sort of curiosity that is edge driven, not only learning what and how things
belong, but also building heretofore unimaginable lines of belonging. It is
inquisitiveness untethered to acquisition.

Long ago, Augustine traced curiosity to the eyes—the eyes that look,
observe, search, and explore. Later, Descartes diffused it in the blood,
suggesting that "animal spirits" shuttle about and suspend our attention
on novel perceptual impressions, while naturalist Charles Bonnet and
empiricist Étienne Bonnot de Condillac rerooted it in the organ of touch,
especially the human finger or the elephant's trunk.[14] It was not until the
nineteenth century, with the advent of phrenology, that curiosity found its
place in the brain. Appearing under the term "marvelousness" or "wonder,"
curiosity was located by phrenologists H. Lundie and George Combe on
either side of the parietal lobe, where it functioned not as an intellectual
faculty but instead as a "superior" or "affective" sentiment.[15] Of course,
with the development of neuroscience, contemporary scientists have used

fMRI to show that the brain is not simply a conglomeration of sections, each devoted to one specific task, but rather it is a network of information channels, the flexibility of which preconditions function.[16] On this model, while certain signals of curious behavior are traceable to the hippocampus or lateral parietal lobe, curiosity itself is necessarily diffused.[17] It cannot be one thing, located in one place, practiced by one person, in the pursuit of one piece of information. But if not one, then what? What would it mean to rethink curiosity from a relational or even a network perspective? To take up the art of forming and fashioning the concept of curiosity anew?

Imagine a curiosity that is more social and praxiological than it is individual, intellectual, and acquisitional. Imagine a curiosity that aims less to know X, to find out X, or to cognize X than to make connections, build constellations, find links, and follow threads. Imagine a curiosity that is collective and interconnective, functioning within a webbed network of relations between knowers, methods of knowing, and knowledges. While the traditional view of curiosity locates its practice in the individual act of collecting information, tracking down answers, or imagining something new, a relational and network view of curiosity would expressly embed inquisitive affects, practices, and architectures in deeply imbricated (eco) systems and (eco)cultures.[18] Curiosity from this vantage point would be less like the eye than the light itself. It is an intriguing suggestion at the very least, but is there historical precedent for such a concept? In short, yes; there is a markedly robust precedence for this view. While certain historical critiques of curiosity precisely castigate it for breaking connection and creating isolation—recall early modernist Blaise Pascal writing, "I see the depths of my pride, curiosity, concupiscence. There is no link [*rapport*] between me and God"—on the whole, there is nevertheless a steady trace of thinking curiosity as a relational practice of connection between knowers, methods of knowing, and (un)knowns.[19] Here we follow that trace, culling critical characteristics of curiosity as a practice of connection.

For ancient thinkers, curiosity has the capacity to produce two essential connections: 1) a connection between the inquirer and their inner selves and 2) a connection between one object of inquiry and another. Plutarch makes good fun when he surmises that the chatterbox—close cousin of the curious busybody—has an atypical anatomy: where a conduit between ear and soul should appear, there is instead one between ear and tongue.[20] A rogue eustachian tube straight into the gullet! The words just tumble in and tumble out, via blustery byways, never landing in the sweet earth of the soul to germinate and bear fruit. Plutarch's word for conduit here is *suntetraino* (Gk. *sun*: with, together; *tetraino*: to perforate), which refers to boring a connective passage or channel between things. When righted, curiosity rebores proper channels, especially between the knower and themself. When the ancients like Plutarch and others recommend "turning back the gaze," they do not mean to suggest replacing one thing with another, as if to reduce the self to yet another object of knowledge; rather, they enjoin attending to the tension and the distance that separates us from ourselves and our own aims. Their call is not to knowledge, then, but rather to "attention," from the Latin *tendere*, meaning to stretch between things.[21] But self-relation is not the only connection that curiosity curates. Medieval educational theorist Juan Luis Vives adds the relation between (un)knowns. With "curiosity," he writes "some things . . . follow from the finding of others, just as when the beginning of a thread is secured, it is found to be connected with another set of things quite different from those which were being examined."[22] This is the pleasure of curiosity, the tautness of a thread in one's hand, leading one onward.

Some modern philosophers, likewise, think curiosity as an activity of relation, not so much of the self's relation to the self, but of the self's relation to objects and those objects' relations to each other. For Thomas Hobbes, curiosity is an interest in causal relations, shuttling from an effect back to all possible causes or from a cause forward to all its possible effects.[23] More robustly, Jean-Jacques Rousseau conceives of curiosity as governed by existing connections between knower and (un)known, as well as guided by existing connections between (un)knowns. In *Émile*, his treatise on education, Rousseau deftly distinguishes between an "ardor for knowledge" that aims to merely impress, and one that aims to pursue what is of inherent interest or concern to oneself.[24] Natural to every human heart, curiosity extends to anything and everything that is "connected" to an individual inquirer's well-being. Much as Friedrich Nietzsche decries a curiosity that lacks personal investment, insisting that "all great problems demand great love," Rousseau too insists that curiosity must stem from personal investment for it to be authentic.[25] Since, however, ultimately "everything concerns us," or is connected to us, and everything is connected to everything else, such that "each individual object brings forward another," à la Vives, every single thing is a potential object of curiosity. What is required of the human knower is that curiosity start from a personal connection, proceed via the (un)known's connection to other (un)knowns, and constantly reconnect knowledge back to personal concern and social good. "What remains for us to do after having observed all that surrounds us?" Rousseau concludes, but "to convert to our use all of it that we can appropriate to ourselves, and to make use of our curiosity for the advantage of our own well-being."[26]

Building and expanding upon these ancient and modern insights, the pragmatist philosophers of the late nineteenth and early twentieth centuries are exemplary in the depth with which they think curiosity in relation. This is perhaps no surprise given their commitment to embedding knowledge of the world in experiential experiments with that world. William James recommends cultivating curiosity by "stirring up connections" between the knower's nature and the unknown, whether via "personal interest," "pugnacious impulse," or "theoretic curiosity."[27] "Theoretic curiosity," he clarifies, is stimulated by "the rational relations between things"—or in Charles Sanders Peirce's terms, "the very remarkable relations" that demand "rational explanations."[28] Once the connectedness between the (un)knower and (un)known has been primed, connections between the unknown and known can be built. As James memorably puts it, curiosity finds in new facts "an old friend"; it "knits" new information to old information and fresh curiosities to former curiosities.[29] Further developing this account, John Dewey begins by insisting that curiosity itself is not "an accidental isolated possession," nor does it function by "amassing," "accumulating," or "heaping up" disconnected bits of information. It is neither merely nodal nor primarily marked by nodal acquisition. Rather, curiosity springs from "the direct interest of life." And it carries an inkling that whatever one currently knows is never "the whole story"; there is always "more behind" and "more to come." As such, curiosity is an impulse to make the "connections" between things—and ultimately the condition of connectedness itself—perceptible.[30] The pragmatic knower thus follows a thread from the self to the world, discovers the knits between things in that world, and then knits themselves back into that web, that worlding.

Although inconsistent with the traditional view and its current manifestations, there are undeniably resources, scattered across the history of Western philosophy, for thinking curiosity at the edge, as a practice of connection and a promulgator of relation. Interestingly, we do not observe any significant chasms between eras or dramatic disjunctions in descriptive accounts. Instead, we observe, era after era, a deepening of appreciation for the relations that produce and are produced by curiosity. In this narrative, at least four sets of relations surface: the relation of an (un)knower to themselves, of an (un)knower to an (un)known, of (un)knowns to other (un)knowns, and of (un)knowns to (un)knowers. These are the caverns that curiosity spans and the spokes on which it spins. Curiosity draws the externally facing inquirer back into relation with their internal world. Curiosity draws the inquirer out toward things with which they are fundamentally connected and concerned. Curiosity follows, feeling along the factual and functional relations between things in the world. And curiosity reconnects what is learned back to the knower and the knowing community, such that both are improved in its wake. Buried here in history, then, is the material for a new paradigm of *curiosity as edgework*. What are the implications of this relational account for psychology, neuroscience, network theory, and beyond? And what might be the practical results of such a paradigm shift?

EDGEWORK IN ACTION

As connective architectures, inherently defined by pinpoints and interstices, networks are nothing if not nodes and edges, units and relations, at once. While we have held nodes and edges at arm's length up to this point, so as to better illuminate the overemphasis on nodal acquisition in traditional accounts of curiosity and entertain the reconceptualization of curiosity as edgework, realistically a blend of these two viewpoints is required for any network account of curiosity worthy of the name. Curiosity is no game of *Pac-Man*, with its traveling mouth munching power pellets. But nor is it merely a performance of Charlotte's web, confined to nets of affective belonging and communications. But how are nodal acquisition and edgework best balanced? How are they imbricated? How can their contiguous natures be best embraced rather than denied or disavowed? Without the connective tissue of (un)knowers, (un)knowledges, and (un)knowns, the moments of nodal creation and growth are decontextualized, and easily misunderstood as unitary gains. The traditional view not only overemphasizes nodal acquisition as the primary marker of curiosity but also locates the edge in desire, stretched between individual and information. The account we have excavated here privileges the edgework of curiosity, and it locates the node as always already coconstituted and connected by other nodes and edges in multiple networks. Such an account builds upon rather than breaks those connections. Standing in a swirl of words and things, such an account positions relations at the forefront, whether between (un)knowers, (un)knowledges, or (un)knowns.

In this final section, we elucidate this network paradigm for curiosity through two different philosophical systems, one contemporary and the other ancient: the work of Michel Foucault and the tradition of Indigenous philosophy, respectively. Each in their own way illuminates the potential significance of curiosity reconceptualized as a practice of connection, within a complex system of overlapping networks.

Foucault is a twentieth-century thinker of power and, specifically, the power to resist established ways of thinking and doing. For him, curiosity is central to that resistance insofar as it reenvisions and revamps inherited pathways.[31] Musing on the perennial suspicion of mass media—as a dystopian deluge of details that dulls the mind and deadens the will—Foucault instead insists upon the proliferation of information and its inherent importance for curiosity. The exponential growth of information networks, whatever the value of the information itself, is crucial for the critical capacity to think, do, and be otherwise. Put quite simply, information networks precondition curiosity. "I dream of a new age of *curiosity*," he writes. "We have the technical means; the desire is there; there is an infinity of things to know; the people capable of doing such work exist. So what is our problem? Too little: channels of communication that are too narrow, almost monopolistic, inadequate."[32] For Foucault, curiosity is a *stance* more than anything else; it is a passion for the present, and an irreverence for long-sedimented values and structures; and it is a readiness to be surprised and a determination to try a different point of view. For curiosity to function, the chains that constrain information must be broken and new information channels must be born in their place. What is needed is a burgeoning of bridges, an explosion of edgework. "We must increase the possibility of movement backward and forward," along ever more differentiating networks.[33] It is only via this growth pattern—of multiplying links, relations, and connections—that newly curious configurations can be conceived, new architectures erected, and new affinities sewn together. Curiosity is a rogue shuttling across networks, a randomized walk, a lunging across lattices that has the capacity to upend it all.

For Foucault, this work of curiosity is not simply intellectual or ideational in nature. He prefers never to speak of power or knowledge in isolation but instead always of power-knowledge, the hyphen signifying that information is always material and the embodied is always theoretical. Burgeoning information networks are also fast-proliferating webs of possibility for affects and actions, sedimentations of power and acts of resistance. In a late interview, Foucault discusses this lived resistance in network nomenclature. Queer people inherit a set of practices from the cisheteronormative world that includes settled ways of speaking, thinking, feeling, and relating both sexually and romantically. That is, there are well-worn pathways, deeply etched network architectures that typically define the romantic sagas of men and women. In the face of those established networks, the task of queer people is to "escape . . . [these] readymade formulas" and "invent" something still "improbable": "affective intensities" and "way[s] of life" that traverse inherited formulas and institutionalized patterns. That is, queer life requires reconfiguring networks, revamping the ways a graph is walked. To do so, queer people have to think and move in a "slantwise" direction, constantly creating "diagonal lines" in the social network. Moving in these new patterns requires that they "invent . . . the meaning of [that] movement." They must generate fresh forms of signification and sources of significance; they have to tell new stories. Ultimately, given that "what exists is far from filling all possible spaces," the generation of more channels of information creates more chances for resistance and reconfiguration.[34] With more fabric, there can be more fabrication.

The Foucauldian assumption that networks of information precondition ways of thinking, doing, and being has an ancient, rich, and still robust precedent in Indigenous philosophy. Rooted in the wisdom that everything that exists is connected to everything else, Indigenous philosophy foregrounds the vast and complex system of relational networks. While Western philosophy, especially post-Enlightenment, has typically emphasized the individual nodes of knowers and knowns, Indigenous philosophy has consistently contributed to a thinking of the edge. It is not insignificant that the English language is 70 percent nouns, while Potawatomi is 70 percent verbs.[35] Or that Western settlers conceptualize land as private property and commodity capital, while Indigenous peoples understand it as a connective tissue in a larger gift economy. The difference in ethos between *piecemeal* and *of a piece with* could not be more pronounced. In an Indigenous onto-epistemology, one is always coming to know in intimate relationship with other knowers, including not only community members, but also all the components of the earth itself. In *Braiding Sweetgrass*, Potawatomi botanist Robin Wall Kimmerer tells the story of her own Indigenous curiosity. Growing up surrounded by "shoeboxes of seeds and piles of pressed leaves," she knew the plants had chosen her.[36] Declaring a botany major in college, she soon learned to stockpile taxonomic names and functional facts, all while letting her capacities to attend to energetic relationships fall into disuse. It was not until rekindling her connections with Indigenous communities— and specifically Indigenous scientists—that she remembered how "intimacy gives us a different way of seeing."[37] Her scholarship and outreach are now focused on honoring this ray of scientific and social wisdom.

What is perhaps most distinctive about Indigenous philosophy is its imbrication of a relational cosmology with a relational epistemology. At the heart of this worldview is "the eternal convergence of the world within any one thing," writes Carl Mika, such that "one thing is never alone and all things actively construct and compose it." From this perspective of deep holism, a metaphysics of presence is inherently fragmentary and talk of knowing any one thing is "minimally useful."[38] Just as existence is always already embedded within contiguous *existants*, knowledge is always already embedded within local frames. Against a colonial knowledge system of delocalization and abstraction, then, Brian Burkhart argues for a restoration of "intimate knowing kinship relationships."[39] Such relationships dispense with a static emphasis on truth acquisition and focus instead on the dynamics of meaning making, traditionally supported through ceremony and storytelling. In such settings, it is the individual in the community and the mind united with the body that precondition deep understanding. As Thomas Norton-Smith expresses it, in Indigenous philosophy "relatedness [is] a world-ordering principle." As such, knowledge is not properly propositional but instead procedural; it is less concerned with knowing what than with knowing how. And its wisdom lies in "sharing" more than "stating." Asking questions, then, involves seeking out existing patterns and emerging connections through what is "lived" and what is "embodied."[40] Curiosity inescapably arises and is satisfied in relation.

Although traditional accounts of curiosity in Western philosophy typically stress its function in the nodal acquisition of knowledge by individual, independent knowers, a network approach necessarily submerges the node in the network and rebalances attention on the energy of the edge. While we have traced an underappreciated history of thinking curiosity as a practice of connection, or as edgework, across Western philosophy, we have also deepened that account by considering it in action within the political philosophy of Foucault and the epistemology of Indigenous philosophy. Still, more than one line of inquiry remains open. Within the Foucauldian and Indigenous accounts, for example, what is the precise interplay between information networks and social networks, or thought networks and species networks? And what is curiosity's role in social change and environmental healing, respectively? More broadly, how does a network reconfiguration of curiosity challenge not only contemporary philosophies of curiosity but also recent accounts of curiosity in psychology and neuroscience?

*   *   *

Recall that while psychology focuses on curiosity as a mental state or a personality trait with corresponding behaviors, neuroscience aims to identify the neural substrates of curiosity. Within contemporary accounts in both fields, curiosity is typically understood as an organism's motivation to acquire and accumulate knowledge. This motivation is characterized, by turns, as the need to fill a gnawing information gap or as the evolutionary capacity to secure information necessary for survival and flourishing. In many cases, such curiosity is tested through assigned tasks of manipulation or answers to trivia questions. Reconceptualizing curiosity as a practice of epistemic connection and the curious subject as essentially embedded in relations prompts a deepening analysis of curiosity not only in philosophy but also in social psychology and cognitive neuroscience. Within this frame, curiosity is less of an abstract affect, or state or trait, and more of a practice, always in motion, stretched across informational and social networks. Against a theoretical framework that functions to individualize and isolate—whether the inquirer or the information—a relational, network approach invites a renewed investment in ecological validity, thinking and rethinking curiosity in situ. Such a perspective is promising insofar as it reroots an analysis of curiosity in the node-edge relation, whether considered via neuroscience, psychology, and philosophy, or transdisciplinary conversations between them.

Let us return for a moment to physician and natural historian Olaus Wormius. The intrepid investigator that he was, he traipsed around the globe and collected—indeed acquired—untold troves of artifacts. A tortoise shell here, an armadillo and elk horn there, shawls from one village and armor from another. From coins and fossils to original artwork and Indigenous artifacts, he brought everything back to Copenhagen. There he crafted his Museum Wormianum, a cabinet of curiosities hardly rivaled in the seventeenth century. It is easy to think of Ole Worm as a solitary *curiosus* and his collection as a stockpile of curios. But each of these objects was extracted from a network of relations, social systems, and ecologies in which they played a material as much as a symbolic role. Not only were these curios deeply edged, like starfish, but Ole Worm's own curiosity was also a cog in the colonial machinery, his interests like so many arms and legs of an imperial natural history. Curiosity and curios are never in fact solitary. But this begs the question, How does one honor (or dishonor) the relations in which one's curiosity and curios are embedded? What edges does one feel for? And to which does one sustain fealty? What might the Museum Wormianum look like apart from the colonial apparatus in which it took meaning and shape? Ole Worm himself was no recluse. Often accompanied by his wife, Dorothy (who invented the term *tangent*) and his pet auk (a penguin look-alike), he was the sort of person who remained in Copenhagen during the Black Plague to care for the sick—even at the cost of his own life. What is a curiosity grounded in tangential belonging—one that grants the radical interdependence of our lives and our insights?

When one is curious, how does one catch and craft a thread? It is a long-standing hunch in English literature, from children's tales to adult science fiction, that in fact all the world is threaded together. If only we could see it! We might, as Irene does in *The Princess and the Goblin*, position a thread between our finger and thumb, and journey out into the world, following it wherever it may lead, and ultimately find the cipher of being, truth, and belonging. Or like Mark in *The Changeling*, we might pull on the threads of varying lengths and colors that hang about our world and, from them, gain different kinds of power, as if the way in which things belong together were a source point for cosmic energy and insight. Writer Henry James, in "The Art of Fiction," memorably observes, "Experience is never limited and it is never complete; it is an immense sensibility, a kind of huge spider-web, of the finest silken threads."[41] The work of fiction—or of crafting sense through experience—is a work of threading. How does one follow the threads? And by what force does one weave them in and out so as to craft the fabric of a curious life? Edgework invites us to ask these questions in a way that nodal acquisition never could.

At the close of the medieval period, philosopher and theologian Desiderius Erasmus bemoaned the irreverent curiosity that feverishly ferrets out the secrets of nature; especially galled by the hubris of astronomy, he called the "measure of the stars" sheer "madness."[42] By the late twentieth century, however, the tables had turned. In his *History of Astronomy*, philosopher and economist Adam Smith not only praised the practice of a curiosity trained on extraterrestrial sites but also remarked that "curious attention . . . labors to connect one thing to another."[43] Without having to turn to the network of stars overhead, we turn to the network of neurons within us and the notions that traverse us. And we investigate what those connections might teach us about the architectures that characterize curious minds. In the next chapter, we therefore explore curiosity less as a principle of knowledge acquisition than as a practice of knowledge network growth.

3 THE NETWORK PARADIGM

> Curiosity is not an accidental isolated possession; it is a necessary consequence of the fact that an experience is a moving, changing thing, involving all kinds of connections with other things. Curiosity is but the tendency to make these conditions perceptible.
> —John Dewey, *Democracy and Education* (1916)

As we look up from the rich annals of history, we reengage with our surroundings. Where were we? Ah yes. In Copenhagen, watching Ole Worm stoop to pick up an artifact for the Museum Wormianum. Stretching our legs gingerly, we retreat from the deep recesses of darkened pages into the fresh living tissue of our minds. From this most familiar of vantage points, we look out at the world. We find—oddly—that the clock ticks more loudly than before, the air takes interest in bending the little hairs on our arms more curiously than before, and the light streams through the window more colorfully than before. Colors. The colors of light. Many changes in the colors of light.

How does the same light, here just a moment ago, look so different now? Scientist and writer Johann Wolfgang von Goethe offers a thoughtful account in his 1810 *Theory of Colours*, which composer Ludwig van Beethoven kindly admits to be "important," unlike his other "insipid" writings.[1] In the treatise, Goethe sought to trace "the phenomena of colours to their first sources, to the circumstances under which they appear and are." Toward the end of the book, he notes that the color curator and conservationist (the scientist), after cohabiting with a philosopher as we have just done, acquires a different relationship with the world. "The investigator of nature cannot be required to be a philosopher, but it is expected that he should so far have attained the habit of philosophizing, as to distinguish himself essentially from the world, in order to associate himself with it again in a higher sense."[2] We have distinguished ourselves from the world in successfully extricating the unitary knower, acquiring unitary knowledges, from our inquiry into curiosity. In so doing, we have gained a connective prism, an instrument of perception and measurement, linking ourselves differently to the world around us. Holding the prism to our eyes, we stand tiptoe at the edge of our seeing orbs, . . . the light of our curious thoughts bending in colors toward and into the light of the world's adjoining sphere. If curious thought is connective, then a science of curiosity must be a science of connection. And not of a single connection but instead of *connection* plural . . . of many-colored connections.

With new eyes, we revisit the foundations of a science of curiosity. What does it mean to connect? Where do we see connections in Nature? How have they been studied in the past? What instruments, models, and theories have proven useful? Does there exist a scientific discipline that can accurately be called a science of connection? If so, might that discipline prove useful in the study of curiosity? Let's start at the beginning, where a single connection links two things. Once grasped, the diminutive artifact connects Ole's pointer finger with his opposable thumb, which he shares with the loris and lemur, but not with *leontopithecus*, that fluffy golden lion of the trees. Is this the sort of connection we are looking for? One that connects by touch, like snow resting on the ground, or one that connects by physical abutment, like my back leaning against the base of a tree as I watch the wind tussle the leaves and the leaves tussle the squirrels? Or do we seek a connection that tethers, like a rope or string, or perhaps a connection that conducts, like a duct or vein? The sensing pad of Ole's thumb sits atop the bony substrate of the distal phalanx, which connects to the proximal phalanx by the collateral ligament. The ligament tethers the two bones to one another and conducts force, thereby ensuring that the distal phalanx never spins like a top as we twiddle our thumbs. Lucky. Is this the sort of connection we are looking for? Or do we seek a connection that generates, like an umbilical cord linking the giver to the receiver to support life, or perhaps a connection that engenders, like a spinal or ventral nerve cord that gives rise to the tickle of senses? Along Ole's distal phalanx runs the proper volar digital nerve that transmits the artifact's tactile features up to the brain, where the item is placed within his broader collection by the mental manipulations of consideration, comparison, and contrast. The connection of the new to the old, and vice versa, generates understanding and engenders realization. Connecting one idea to another, one concept to another, changes not just how Ole thinks about each item separately but also how he thinks about the broader structure of the tree of life.

Connections of touch or physical abutment, that tether or conduct, to generate or engender, can link two things. But is two enough for a connectional account of curiosity? Or do we need to understand how to connect more than two? Two is enough for life and death, for light and dark, and more generally for any line with dual ends. A continuum, a pair of opposites, black-and-white binaries or simple boundaries, all need only a dyadic connection. Where we start running into trouble is when we think about sets. How do we parse the notion of connection among three particles (electrons, protons, and neutrons) or three postulates (Newton's laws of motion), among four particles (gauge bosons) or four postulates (Maxwell's equations of electromagnetism), among five Platonic solids (tetrahedron, cube, octahedron, dodecahedron, and icosahedron) or six quarks (up, down, top, bottom, strange, and charm). What is this connectional notion of belonging, of community, or of family resemblance, as Austrian British philosopher Ludwig Wittgenstein might have put it? Perhaps the problem (and its solution) will become more evident if we move from the abstraction of physics to the tangibility of biology. Let's consider one head, two ears, three ethmoid sinuses, four appendages, five fingers . . . Ah yes, Ole's sensing pads, each sitting atop its own bony substrate of a distal phalanx. These five are connected as one: a hand that allows him to grasp the artifact. The palm serves as a generic, indiscriminate connector among all elements of the set, not simply among element pairs. Here then is a different sort of connection, a set-type connection, for which we as scientists use a different word. Whereas a dyadic relation is typically referred to as an *edge* in science, a set relation is typically referred to as a *hyperedge*. In our notepads, we draw a hyperedge as a blob-like outline that reaches and encircles each element of the set, visually reminiscent of the bodily outline of an amoeba extending fingerlike projections of protoplasm or the bowed outline of a toddler's rounded hand with pudgy phalanges spread out so as to be traced on pink construction paper. Both the edge and the hyperedge are unitary, each reflecting a single type of relation in a single connection.

Connections of a dyadic or multiadic linkage allow us to understand a single relation among any number of things. But is a single connection—no matter how large or multipronged—enough for a connectional account of curiosity? Or do we need to understand how to link connections to one another? As a simple thought experiment, imagine a triangular shape. In nature, we see these shapes in beech nuts, trillium, trefoil, spiderwort flowers, oxalis leaves, the angle shades moth, deltoid leaves, centric diatoms, mountain silhouettes, cucumber flesh, tail fins, and orchid blossoms. Moving from nature to mind, we can decompose the triangular shape into its compositional elements: three beams connected at the joints by hinges to form a triangular frame. By transforming the shape into a frame, we obtain a more elemental structure that allows us to think differently about concepts and their linkages. Consider, for example, a triangular frame stitching together the concepts of light. At the top junction of the triangular frame sits green light, at the bottom-left junction sits orange light, and at the bottom-right junction sits purple light. The edge of the frame connecting green to orange light reflects the relation "shares yellow light." The edge connecting orange to purple light reflects the relation "shares red light." And the edge connecting purple to green light reflects the relation "shares blue light." In all, we have three distinct concepts and three distinct edges, and we know from experience that these ingredients form the shape of a triangle.

Now imagine the triangular frame again, but this time with one edge missing. In nature, we see this structure in a branching twig, a forking path, a bifurcating vessel, a splitting fiber, a flying skein, or a diverging river. Moving from nature to mind, we might see this structure in the relations

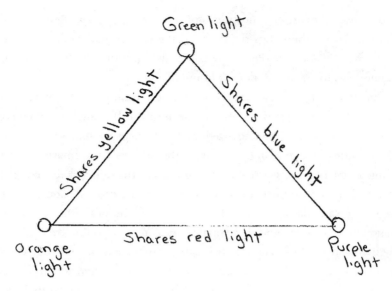

Figure 3.1

A triangular frame stitching together the concepts of light. At the top sits green light, at the bottom-left sits orange light, and at the bottom-right sits purple light. The edge connecting green to orange light reflects the relation "shares yellow light." The edge connecting orange to purple light reflects the relation "shares red light." And the edge connecting purple to green light reflects the relation "shares blue light."

among three concepts, two of which may or may not be related depending on the context. For instance, the concept of red is linked to the concept of blood by the relation "shared color," and the concept of blood is linked to the concept of health by the relation "shared association with strength and vitality." Yet as avid readers of medieval adventure stories know, red and health may or may not be associated with one another depending on the context. When the red of the blood remains inside the body, we associate red with health, as in the flush of action or the blush of inaction. In contrast, when the red of the blood breaks through to the outside of the body, we associate red with death, as in a mortal wound. The wedge shape, unlike the triangle, here provides the potential for contextual flexibility as much as for stored energy and the spring of possibility; like the wishbone (*furcula*: fused clavicles of a bird), the wedge shape captures the energy of the downward thrust of conceptual contraction only to spring back in the upstroke, lifting the wings for flight in the conceptual expansion of the mind. Note that the wedge shape could also indicate that the third link does not exist or is simply not yet known.

Triangles and wedges are two common examples of connectional motifs. When we are curious, we may seek (or find) triangles among concepts, or we may seek (or find) wedges. Or perhaps we may seek to transform a wedge into a triangle by filling the apparent information gap by adding a new edge between the trailing lines of the flying skein. But need our minds be constrained to only trade in triadic motifs? Could we not imagine rectangular, pentagonal, or hexagonal frames? Or heptagonal, octagonal, nonagonal, and decagonal frames? Yes, we could. We could also imagine each of those frames with one or more edges removed. Or we could imagine each of those frames with one or more edges added between nonadjacent junctions, iteratively transforming the polygon into an Anishinaabe dreamcatcher waving in the wind. As we grow, develop, and learn, it seems natural that we might build not just one shape of connections between concepts but instead many shapes—some stitched together, some isolated, and some socially distanced waiting for the time and freedom to draw closer. In some areas of our lives, perhaps we build in grids under the auspices of a comprehensive formal education and then circumscribe our knowledge of a rigorous discipline as if placing a smooth rim about the string mesh of a tennis racquet face. In other areas of our lives, perhaps we build in mixtures of geometric tiling and tessellation reminiscent of quasicrystals (p-type, f-type, or icosahedrite), continuously filling our available conceptual space, but in a way that lacks translational symmetry. Probably in most of our lives, though, we build in architectures far from grids, meshes, and tessellations, and more akin to river networks, the mycelium of fungi, dendritic arbors, and the rhizomes of sweetgrass. How do we begin to understand how a particular flow of thought makes the river network of a mind? Why and how does the network of one mind differ from that of another? How do we dissect, detect, discern, discover, and describe the connectional patterns we build as we lay down edgework in curious communion with our world?

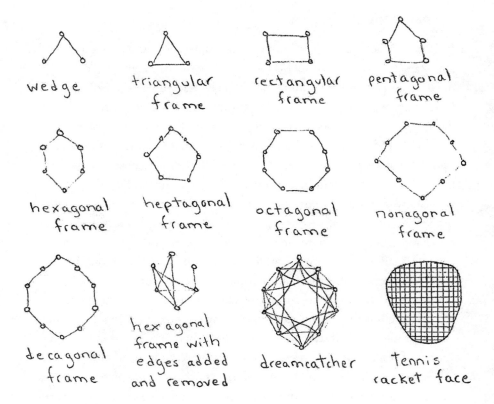

wedge

triangular
frame

rectangular
frame

pentagonal
frame

hexagonal
frame

heptagonal
frame

octagonal
frame

nonagonal
frame

decagonal
frame

hexagonal
frame with
edges added
and removed

dreamcatcher

Tennis
racket face

Figure 3.2

Connectional motifs. The connectional structure of a wedge (three nodes, two edges), triangular frame (three nodes, three edges), rectangular frame (four nodes, four edges), pentagonal frame (five nodes, five edges), hexagonal frame (six nodes, six edges), heptagonal
frame (seven nodes, seven edges), octagonal frame (eight nodes, eight edges), nonagonal
frame (nine nodes, nine edges), and decagonal frame (ten nodes, ten edges). Note that any
of these frames can have edges added or edges removed, or both; as an example, here we
show a hexagonal frame with one edge removed and five edges added. If we take the decagonal frame and add many edges, we arrive at a connectional structure that looks a bit like a
dreamcatcher. Sometimes perhaps we build connectional structures that are less dreamlike
and more grid-like, similar to this tennis racket face.

PAINTING NETWORKS

In Robert Macfarlane's *Landmarks*, a marvelous book about language and landscape, he recalls inviting his mentor, Roger Deakin, an English writer, documentarian, and environmentalist, to give a lecture at the University of Cambridge regarding his work on UK waterways. Macfarlane recalls Roger departing from the expectation of rigorous logic that is characteristic of the typical Oxbridge seminar. Water—which was Roger's subject—"doesn't do rigor in that sense, and neither did Roger," Macfarlane observes, as if the blue of water were definitively not the blue of Cambridge. "For Roger, water flowed fast and wildly through culture: it was protean, it was 'slip-shape' . . . and so that was how he followed it, slipshod and shipshape at once, moving from a word here to an idea there, pursuing water's influence, too fast for his notes or audience to keep up with, joining his . . . watery subjects by means of an invisible network of tunnels and drains."[3] Roger's mind followed his subject of curiosity (water) by tracing out its network architecture (waterways) in the way he spoke (flowing, tunneling, and draining). It's as if water saw its own comely face reflected in the liquid surface of Roger's words. This type of reflection is reminiscent of the notion of *ductus* in ancient and medieval rhetoric: "the 'route' in which one moves through a composition, a path marked by the 'modes' or 'colors' of its varying parts, . . . the basic direction of a work, which flows along, like water in an aqueduct, through whatever kinds of construction it encounters on its way," writes Mary Carruthers, a notable scholar of medieval rhetoric and aesthetics. "Indeed, *ductus* insists upon movement, the con*duct* of a thinking mind on its way through a composition."[4] What is our contemporary *ductus*? Through what routes, paths, and aqueducts do we move, and in what modes and colors? How does the mind track and trace through the network structures of knowledge, either in solitude or in society? And what are those architectures that we trace? For river networks, we can draw lines of passage or paint veins of voyage on a two-dimensional bit of paper. But how, and upon what substance, do we paint the networks of the mind?

In seeking answers, we naturally turn to the formal science of connection. Network science is the study of systems that are composed of many units linked in complex patterns of connection. It provides a mathematical language in which to describe the architecture of those patterns and to explain the functions supported by that architecture.[5] A relatively young field of inquiry, network science saw its first formal institutes around the year 2000, and its first undergraduate and graduate degrees about ten years later. Of critical importance to the missions of research funding bodies across the world, network science has—in a remarkably short time—secured fundamental advances in biology, ecology, neuroscience, social science, technology, and engineering, among others. Each investigation begins with the reductionist approach of delineating the units of the system, which are referred to as network nodes, and the relations between those units, which are referred to as network edges. Then we doff the reductionist's hat (which I always envision as a beanie—in which one counts beans) and don the constructionist's hat (naturally, Sherlock Holmes's deerstalker—in which one connects clues in woven tweed). Appropriately attired, we take up our palette and brushes to paint upon the network scientist's typical canvas: the crisscrossing fibers of a table composed of N rows and N columns, where N is the number of units in the system. At the intersection of the i-th row and j-th column sits a number indicating the strength with which unit i is connected to unit j. To know which intersections to fill and with what numbers, we turn our eyes to the real system, notice each point of identity and each stroke of connection, and then turn back to the canvas to attempt an accurate portrayal. Like any painting, the depiction we produce does not live and breathe as the true system might but instead it represents and it encodes. In painting the network, we construct a summary or model from which we seek to derive understanding.

Understanding arises from a careful study of the network's architecture, which is peculiar to each system. The number of different connection patterns that we could hypothetically observe in the world around us is given by the number of unique ways in which the table can be filled. As a simple first example, let's consider the system of knowledge. Imagine all the bits of knowledge in the world; that number is N, the tally of rows—or

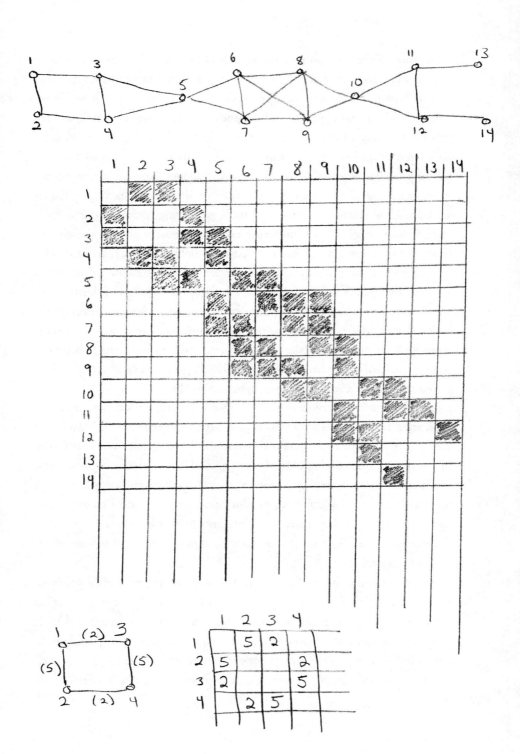

equivalently, the count of columns. Now imagine the many ways in which those bits of knowledge could be related to one another—the many ways to color in the elements of the table. The cardinality of these ways grows, compounds, and multiplies as we consider the idiosyncrasies of minds and the distinctiveness of times, as unknowns become knowns in the process of discovery, or knowns become unknowns in the processes of falsification or forgetting. The possibilities feel boundless. How do we wrangle those possibilities into principles? Or categorize phenotypes into classes? Or catalog patterns onto cards? Network science offers a set of statistical measures that we can use to quantitatively and precisely describe any observed connectivity pattern. Such measures allow us not only to state, for instance, that the iron lattice architecture of the Eiffel Tower is different from the steel lattice architecture of a common transmission tower but also to state the manner and extent of that difference. Similarly, such measures allow one to describe the patterns of movement among ideas, such as in the storyline of novelist Georges Perec's *Life: A User's Manual*, which traces a knight's tour (the movement of a knight on a chessboard) across the ten-by-ten grid (ten floors, with ten rooms per floor) of an apartment building on 11 rue Simon-Crubellier, . . . or as in the narrative of novelist Julio Cortázar's *Hopscotch*, either read in the first order (chapters 1 through 56 in numerical sequence) or the second (chapters 73-1-2-116-3-84-4-71-5-81-74-6-7-8-

---

Figure 3.3

Painting networks. At the top, we show a network with fourteen nodes and twenty edges. Below the network we show a table composed of N rows and N columns where N is the number of nodes (in this case, fourteen). The intersection of the first row and second column is colored in because node one is connected to node two by an edge; similarly, the intersection of the second row and first column is colored in because node two is connected to node one by that same edge. For each edge in the network, we can fill in the corresponding elements of the table. The table as filled in here perfectly encodes the information in the network above it. Note that this example contains edges that either do or do not exist. Some networks, however, have edges that are assigned numbers to indicate not just their existence but also their strength. In the bottom of the figure, we show a small network with four nodes and four weighted edges, and to the right of the small network, we show the network scientist's canvas with elements in the table not just colored in but also filled with the number reflecting the edge's weight.

93-68 and so on). Whether the network of interest is primarily a structure upon which a process can occur (a lattice tower balancing forces to attain height, sustain rigidity, and retain stability) or the outcome of a process upon a structure (the connective tissue of a story constructed by a walk through a conceptual space of characters, events, and locations), its architecture is measurable, calculable, and tabulatable.

One network feature that is particularly interesting to study is local clustering. In a simple network where edges either exist (a one in the table) or do not exist (a zero in the table), the clustering coefficient can be defined as the ratio of the number of connected triangles (three nodes and three edges) to the number of connected triples (three nodes and two edges). Intuitively, a triangular mesh in two dimensions has a high clustering coefficient (many triangles) whereas a hub-and-spoke architecture or star has a low clustering coefficient (zero triangles). The effective closure of triangles supports the confluence of content, current, or comestibles that may be traveling along the network edges. To see how this works, let's examine a copy of prelate and scholar Isidore of Seville's *Etymologiarum*, from the third quarter of the twelfth century, housed in the Bibliothèque Municipale in Douai, France.[6] Richly illuminated, the book opens with a religious poem written on a triangular grid. The page is cut into eight square frames; four span the left side of the page from top to bottom and an adjoining four span the right side of the page. In the center of each square originate eight tubes that emanate radially outward, octisecting the square into triangles. The words of the poem are written around each bit of inner tubing and outer framing, filling the full triangular grid with intersecting phrases, thus inviting the reader to follow the writer's thoughts in a plethora of directions. (An adventure of the choose-your-own-*ductus* variety.) The triangles allow us to arrive at any given junction along many different paths of phrase, bringing with us different contexts depending on the traversal we have chosen. (The geometry of this piece is similar to that of another in the Vatican library and falls under the broad genre of labyrinth poems, which spans from ancient times to today.)[7] Here, triangles guide the fruition of the mind. But we need not stop at the abstract. True, triangles are relevant to the upper echelons of the mind, but they are also relevant all the way

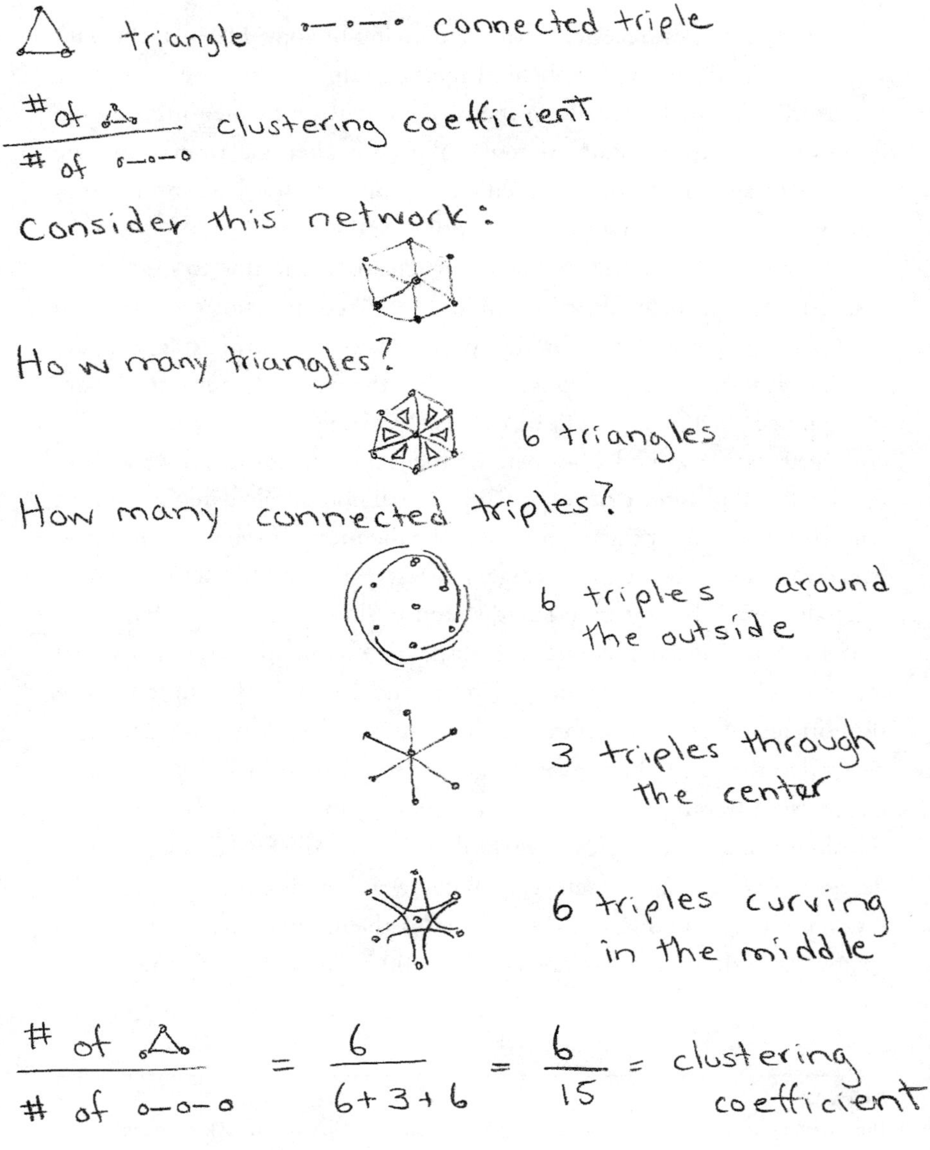

△ triangle   o—o—o connected triple

$$\frac{\text{\# of } \triangle_o}{\text{\# of } o\text{—}o\text{—}o} \quad \text{clustering coefficient}$$

Consider this network:

How many triangles?

6 triangles

How many connected triples?

6 triples around the outside

3 triples through the center

6 triples curving in the middle

$$\frac{\text{\# of } \triangle_o}{\text{\# of } o\text{—}o\text{—}o} = \frac{6}{6+3+6} = \frac{6}{15} = \text{clustering coefficient}$$

Figure 3.4

Pictorial depiction of the clustering coefficient. A triangle is composed of three nodes and three edges; a connected triple is composed of three nodes and two edges. The clustering coefficient of an unweighted network is given by the ratio of the number of triangles to the number of connected triples. As an example, consider the network shown here. How many triangles does it have? Six triangles, as indicated by small pie shapes. How many connected triples does it have? There are six triples around the outside, three triples through the center, and six triples curving in the middle. Hence the clustering coefficient of the innermost node is six (triangles) divided by (six plus three plus six equals) fifteen connected triples, or six-fifteenths.

down to the deepest recesses of the underlying biology. Here, in the dark wetter spaces of cells pulsing behind their curtain of bone, triangles guide the fruition of the brain. Along the physical paths by which information is transmitted in the brain, triangles allow for alternate routes, and are thought to support flexible and efficient propagation of the information necessary for human cognition, perception, and action.[8]

A second network feature that is particularly interesting to study is the shortest path. In the medieval poem just described, the shortest path from the upper-left corner (initial node) to the lower-right corner (target node) requires us to traverse eight phrases (edges); there are, of course by design, many longer paths that could be traversed starting at the initial node and ending at the target node, but here we focus on the shortest. If we were to enumerate all possible pairs of initial and final nodes, calculate the shortest path between each of them, and then take the mean of the resultant numbers, we would obtain the average shortest path length. Networks with generally small shortest path lengths tend to do so by having a few edges that serve as shortcuts, connecting disparate sectors of the network that are otherwise largely disconnected from one another. The shortcuts can support the effective transport and distribution of nutrients in vasculature networks, the swift sharing of ideas between people in social networks, or the rapid leaping of the mind between ideas in concept networks. As a simple example of such a leap, we think of Xu Yin, the Five Dynasties poet, describing *mi se* (a jade green celadon color) as "carving the light from the moon to dye the mountain stream."[9] To link the moon to the earth requires a conceptual shortcut across the vast 238,900 miles between them, yet that

---

Figure 3.5

The structure of the religious poem written on a triangular grid in the copy of Isidore of Seville's *Etymologiarum* housed in the Bibliothèque Municipale in Douai, France. The page is cut into eight square frames; four span the left side of the page from top to bottom and an adjoining four span the right side of the page. In the center of each square originate eight tubes that emanate radially outward, octisecting the square into triangles. The words of the poem are written around each bit of inner tubing and outer framing, filling the full triangular grid with intersecting phrases.

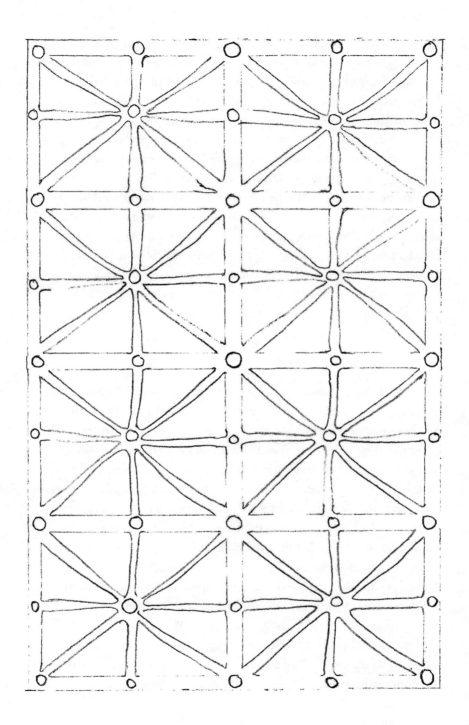

shortcut can be made in the mind by tracing a ray of moonlight. Similarly, to link light to dye requires a conceptual shortcut across the vast difference in spatial scales between a photon (the element of light) and a string of fifty atoms (the typical size of a dye molecule), yet that shortcut can be made in the mind by noticing the close relation between light absorbed and light reflected. The leap is a shortcut in the network landscape that brings two ideas closer than they might otherwise appear. Here then, intuitively, shortcuts guide the leaping of the mind. But again we need not stop at the abstract. Shortcuts are just as relevant to the underlying biology, where they serve to guide the physiological leaping of the brain. The physical wires that connect distinct brain areas, resulting in a short average path length, are thought of as highways along which quite distinct types of information can leap to the relevant local regions for integration and processing.[10] Might it be these physical shortcuts in the brain that allow us to grasp the stream, lift it from its bed, and fling it about the moon?

THINKING NETWORKS

The bounding, springing, and vaulting of thought—and the neural activity that supports it—is germane to our curiosity, and to our pursuit of various lines of questioning. But to use network science to understand that leaping curiosity, we must first choose the units of thought. How large is a thought? How expansive or diminutive is a concept? What size stones are we leaping among here? In "Fawn," Nan Shepherd, Scottish nature writer and poet from the early to mid-1900s, observed her experiences when "that thought came leaping like a startled fawn / to lose itself in cover of the wood," and recounted her pursuit "if haply I might meet / the golden flash of that fair thought again."[11] Here a unit of thought is depicted as small, delicate, vulnerable, new, and free. Yet Shepherd's appreciation of the physical movement of the mind extends to other units of thought, portrayed as large, strong, resilient, old, and grounded. In "Winter Branches," she likens thought to the boughs of a tree, noting the way the woody arms "leap up quivering in the vastness" of the sky just as "the thought of humans / leaping from earth's nurture."[12] Yes, some thoughts seem small and others seem

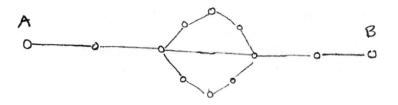

What is the shortest path from A to B?

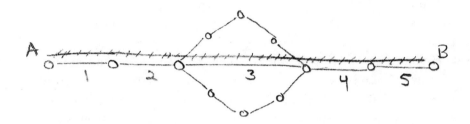

What is a nonshortest path from A to B?

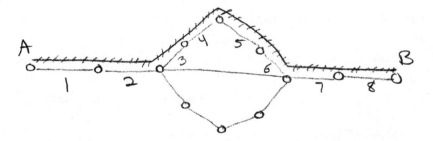

Figure 3.6

Pictorial depiction of the shortest path length. Consider the network shown at the top of the figure. Let's name the leftmost node "A" and the rightmost node "B." What is the shortest path from A to B? We denote the shortest path by the hatch-marked line that passes through the central circle. What is a nonshortest path from A to B? We denote a nonshortest path by the hatch-marked line that passes around the upper part of the central circle.

large. Are the networks of the mind more or less easily painted by network science if we choose small versus large units of thought?

The current evidence in the literature suggests that network science can be flexibly applied across a range of system scales and need not privilege any one scale above others. In its early history, for example, network science was used to build representations of social groups in an effort to understand patterns of social interactions, quantify the influence of a single individual on collective behavior, and predict sociocultural phenomena such as voting patterns or political tumult.[13] Across these efforts, the unit of analysis could vary from small to large: from parts of a person (e.g., brain areas driving social behaviors), to the person as a whole (e.g., an intelligent agent with a particular identity), and then on to groups of many persons (e.g., parties, communities, or countries).[14] The flexibility of network science across units of analysis is also evident in the way it has been used to study the supposed organ of curiosity: the human brain.[15] Network representations of the brain typically begin with a subdivision of the tissue into parcels as small as cells or as large as lobes. Each parcel is chosen because it performs a function, and the parcels' boundaries are chosen according to anatomical markers (the way the tissue is structured) or functional markers (the way the signals are structured).[16] These parcels are then connected with one another using estimates of hardwired connections or of synchronous activity.[17] Across different spatial scales at which parcels are defined, the resultant patterns track a person's openness to experience, creativity, and information-seeking behaviors.[18] In essence then, both society and the brain are multiscale network systems, which—like thought—can have meaningful units defined at different scales.

Although network science can embrace the multiscale nature of thought, it still requires a precise, quantifiable definition of each node. What is a thought, whether small ("three feet tall"), medium ("the size of one's hall closet"), or large ("with grandeur, and drenching, barrel-scorning cataracts, and detonations of fist-clenched hope, and hundreds of cellos")?[19] Ack—like the fawn bounding away through hundreds of trees (or cellos?), boughs (or bows?) at the ready, it is elusive. Perhaps instead we should consider the more concrete fruition of thought in knowledge. What is knowledge? While the *Oxford English Dictionary* provides a quite general account of knowledge as "the sum of what is known," Daniel Webster's 1828 *Dictionary of the English Language* more specifically claims that knowledge is "a clear and certain perception of that which exists, or of truth and fact; the perception of the connection and agreement, or disagreement and repugnancy of our ideas."[20] Knowledge then is a network composed of ideas connected by relationships. Are ideas different from thoughts? Perhaps we will gain yet more traction by considering the precise symbols with which we share and save knowledge: the symbols of language. Language is composed of units that can be defined more precisely than "ideas" and "thoughts." Those units exist over different temporal scales including phonemes (short), syllables (middling), and suprasyllabic objects (long), each of which are represented and processed in different areas of the brain.[21] At the finest level of phonemes, pairs of phonemes are found beside one another with some specific probability, and the set of probabilities defines a network architecture for the language.[22] At the coarser level of phonological word forms, or lexemes, one can similarly construct a network representation by linking lexemes, if they are phonological neighbors of each other in the adult lexicon.[23] The structure of this network is thought to allow us to swiftly and efficiently retrieve word forms from our personal mental lexicons.[24]

Moving from phonemes and lexemes to larger linguistic units, we come to semantic networks, which represent concepts (nodes) and relations between the concepts' meanings (edges).[25] For example, cellos (node) are connected to trees (node) by the concrete relation "is made of wood" (edge) as well as by the more abstract relations "has a sound body," "enjoys being played upon," and "is branched with bows (boughs?)." Early efforts argued that semantic networks were forking structures, without any triangles; connections among nodes were simply determined by class-inclusion relations such as the concept "tree" branching into the smaller concepts of specific tree species.[26] But later work asserted that such a strict, hierarchically branching structure might not account for a majority of concepts.[27]

Today, data continue to mount supporting the notion that semantic networks are not particularly fork-like but instead have a small-world organization (local clustering accompanied by a few shortcuts) and a scale-free organization (most nodes having few connections and a few nodes having many connections).[28] This combination of clustering and connectional heterogeneity is thought to impact how we form and search through our semantic memories.[29] (Ah, there it is!) Notably, distinct types of semantic networks can be built with different relations. Causal links in the network can enable causal inference, causal reasoning, and causal perception, whereas links of simple similarity can reflect the structure of synonyms in a language.[30]

Importantly, because each of us has different experiences, we can each have slightly different appreciations for the similarities or dissimilarities between words; in effect, we may each have different mental networks. To probe people's distinct perceptions of similarity, networks can be built from experiments in which each human volunteer lists a sequence of words where each word is related to the next by meaning.[31] These experiments are referred to as word association or free association tasks, and you may have already met them in a lesser-known work of the English author T. H. White, perhaps best known for his Arthurian novels *Sword and the Stone* and *The Once and Future King*. In White's later novel *Mistress Masham's Repose*, the professor—a curious character—simply cannot remember what he was about to do, so he uses a word association task to solve the mystery:

> An hour later the professor tried to scratch his head, found that the hat was in his way, and wondered why he had put it on. He made several attempts to solve his problem, by free suggestion and self-analysis, finally deciding that he had put it on because he was going out. He therefore went out, and looked at the sky. It did not seem to have any message for him. So he went in again, found a piece of paper, and wrote on it the first word which occurred to him while concentrating on hats. This was tripharium. He tore it up. And tried again, getting ratto, which he thought was probably something to do with Bishop Hanno and the rats. So he tried Hanno and got Windup, tried windup and got cape, tried cape and got ulster. He discovered that he was wearing his

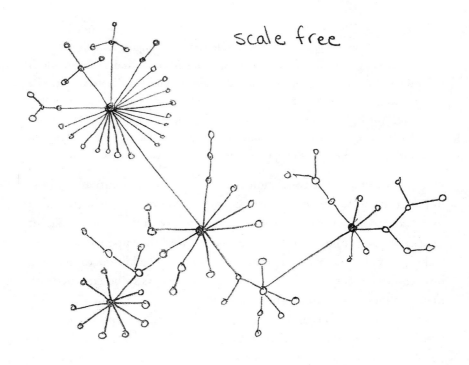

scale free

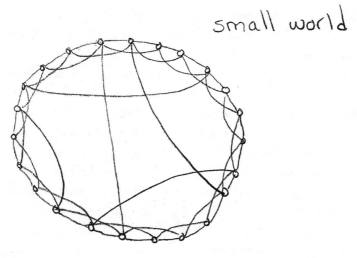

small world

Figure 3.7

Network shapes. Networks can come in many shapes. Two shapes that are particularly relevant to the study of semantic networks are scale-free and small-world. Scale-free networks contain a few nodes with many connections, like the one at the middle of the top-left burst. The majority of nodes have few connections, a couple of connections, or only one connection, like the many nodes at the fringes of bursts. Small-world networks display high local clustering (evident here by the high ratio of triangles to connected triples), accompanied by a few short-cuts (depicted here linking distant sections of the outer ring).

ulster, and was delighted. This was followed by a longish tour through the provinces of Ireland, the Annals of the Four Masters, and so forth, which brought him back to tripharium. He tried this, and got Bloaters, which he connected by now entirely with Sweden, took a short circuit through Gothenburg, Swedenborg, Blake, and Gustavus Adolphus, and suddenly remembered that he was pledged to find Maria. So he balanced the bottle-green bowler still more carefully upon his head, and trotted off to do his duty.[32]

Not only can the word association task remind us of our current purpose in life (as it did for the professor) but it can also provide scientists with a network map of our word-world: which words live in the valleys, adjacent (and connected) to many others, and which words live upon the hilltops, secluded (and disconnected) from most others. In addition to evincing how connected certain words are in a given person's mind, the network map supplies information about distances in conceptual space: which words have their abodes in Boulder, Colorado, versus Miami, Florida, and which spend their days in London versus Lima.

As we move from phonemes to lexemes to words, we steadily walk toward larger units of language and thought. Naturally, after words we come to sentences. While Ole Worm may have preferred the former (binning items into "mineralia," "ligna," and "salia"), I feel as if Virginia Woolf might have invited us to stay awhile in the latter. To her, sentences are the units that defy time ("outlast the Christmas holidays") and yet deserve to be dated ("August, 1928").[33] They are the units that hold life in the balance: a sentence can either kill a mind ("it falls plump to the ground—dead") or create a mind ("explodes and gives birth to all kinds of other ideas").[34] They are the units that comprise the essence of a mind: when meeting someone new on the page, Woolf will first "get the hang of her sentences" as if gauging the author's gait to determine whether they would be good walking companions; she will roll "a sentence or two on my tongue" as if critically sucking a blackcurrant candy to evaluate the depth and verity of its flavor.[35] Sentences can either be "so stereotyped that they chime like bells," or so "undraped" that they evince "a stately and memorable beauty" as a procession of "slightly veiled" bodies whose contours are still visible.[36] Either way, they leave their mark: "sentences roll like drops down a pane, drop collecting drop, but when they reach the bottom, the pane is smeared."[37] Each may contain "minute signals by which a phrase is made to hint, to turn, to live," and then "as phrase joins phrase and one parenthesis after another pours in its tributary, we have a sense of the whole swimming stream."[38] Here again we meet the river of the mind. "The quality," Woolf pens, "is not in the story but above it, not in the things themselves but in their arrangement."[39] The arrangement of things is the network, the manner in which units are connected, the fabric of ideas. We write in networks. We speak in networks. We think in networks.

BUILDING NETWORKS

How do the networks in our mind come to be? Unlike pixies, they cannot just appear of their own accord. Instead, they are built, with every curious action we take. The essence of the nodes gained is determined by the information sought, and the architecture of the network built is determined by the manner of seeking. We may each seek different information, and differently so across our life spans. In young children, information seeking can amount to a heightened attention and focus on objects that are "bright, vivid, startling," while in older adults, information seeking is naturally accompanied by voluntary movements (of the eye or body) to gain more knowledge.[40] One might envision the infant obsessed with the cotton-stuffed ball with colorful velvet patches on the outside and a rich smattering of various sorts of ribbons attached with teeth-resistant stitching, and one might contrast this parochial vision with that of a graduate student entering the Wren Library at Trinity College, Cambridge, seeking a definitive tome on "neurons, networks, and nebulae."[41] Apart from the information sought, curious actions can be characterized by the manner of seeking. These actions—or collectively this practice—of curiosity can differ across individuals, may change with age and cognitive development, and is likely impacted by stress and socioeconomic background as well as prior experience.[42] Intuitively, the practice of curiosity could be impatient or enduring. It could involve seeking completely unknown or vaguely familiar information. It could involve gathering the new information and keeping it logged separately like bits of trivia, or it could involve determining the links between bits of information to fit them into one's existing body of knowledge. While these manners of curiosity are intuitive, they feel nebulous. Might network science allow us to define them more precisely? Could we categorize the practices of curiosity into classes, write down mathematical formulations for their nature, and form generative models for their processes? What sorts of networks might each practice build?

Developing a science of practice invites us to understand the network architecture of knowledge and how it might grow in a single person or across a culture. In place of a picture, we need a movie. In place of a number, we need an equation of dynamics. In place of one network, we need a series of networks, each representing the system at a given point in time. Such network growth models have now been developed for many systems in nature and society, each laying down simple rules for the addition of nodes and the placement of edges. For example, the preferential attachment model of social connection begins with a single edge connecting two nodes and then iteratively adds a single node to the network by linking the new node to some number of existing nodes, with a preference for nodes that are already highly connected.[43] The process naturally aligns with the well-known social dynamics in which already popular people tend to become more popular in the future. Over long periods of time, this growth model tends to form networks with a scale-free organization, whereby most nodes have few connections and a few nodes have many connections. What predictions might a preferential attachment model offer for the practice of curiosity? If the networks of our minds grow with these same tendencies, we would find that each new node of information we add tends to be related to an already well-connected node; in other words, our knowledge of these popular nodes—the topics that we find most intriguing—tends to grow as we learn new associations, new dimensions of sameness and difference, and new likenesses that always lead us from diverse spaces back to our core interests. The resultant structure has pockets of knowledge and ignorance;

the hubs of the former have many connections mapped out in our minds, and the nodes of the latter have few. The preferential attachment model, then, is a particularly useful one for the thinker who cultivates well-defined passions that always draw, call, and compel their attention.

Moving from the large-scale unit of a topic to the smaller-scale units of language, we ask, Are network growth models also relevant for explaining the acquisition of words and word-to-word relations? In one simple model for the growth of semantic networks, new words are added to the network in such a way as to differentiate the connectivity pattern of an existing node; this model operationalizes the intuition that any new concept tends to be a variation on an old one.[44] Its being, its essence, its nature are as framed by its points of relatedness as by its points of unrelatedness. Learning and growing is thus a process of differentiation of meaning, such that the specification of the new simultaneously alters our understanding of the old. This differentiation model generates both small-world and scale-free architectures, consistent with those observed in real word association and synonym networks. Interestingly, this model also suggests a mechanism for the effects of learning history (the age of word acquisition or frequency of word usage) on how humans perform laboratory tasks that demand semantic processing. In a similar spirit, scientists have asked how humans grow specific types of semantic networks such as those evinced by word association tasks. Evidence suggests that a preferential attachment model incorporating word frequency, the number of phonological neighbors, and the connectedness of the new word to words in the learning environment offers a reasonable fit to data acquired from young children as they grow their vocabulary.[45] Such models are generally described as being applicable to all sorts of people. Yet it is also of interest to determine how these models could be adjusted to account for differences *between* people. Do the different ways in which humans build semantic networks arise in part from differences in their personality or brain circuitry, or explain differences in their creativity?[46] Might one's social privileges support—or disadvantages constrain—the ways in which one might build? What other sorts of strictures might determine which edges are possible versus impossible to place in a given network or at a given time?

Accounting for constraints on the network growth process is an important feature of models constructed for real systems. Such constraints are evident across biology, physics, and technology, and they typically align with costs, whether energetic, material, or financial. Efforts to acknowledge and probe constraints on network growth have played a key role in our understanding of development, particularly in neural circuits. Here a common observation is that neurons are more likely to connect to one another if they are close in physical space.[47] A natural model for this process is as follows: distribute nodes randomly in a physical space and then connect nearby node pairs more often than distant node pairs. This model recapitulates certain patterns of mixing observed in neural circuits, whereby highly connected nodes tend to link to sparsely connected ones.[48] Might this same model help us to understand how various factors constrain the growth of knowledge networks? In place of neurons, imagine bits of information; in place of circuit connections, imagine the connections of understanding; and in place of physical constraints, imagine the constraints of social privilege. Distributed differently across people according to history, demographics, and socioeconomic status, social privilege can decrease the cost of growing knowledge, while its absence can increase that cost. The edges we are able to place in our knowledge networks depend upon our access to nature, to conversation, to education, to books, to digital media, and to experiences. Factors that increase the distance between a person and bits of knowledge (or between one bit of knowledge and another) hinder the development of a rich, dense network structure. Social privilege drives growth; marginalization hampers growth. Of course, while marginalized networks are necessarily impoverished by an inequality of access to institutionalized knowledge systems, they can also be differently rich due to the unique constellation of member experiences and alternative knowledge production strategies.

Complementing the constraints of access are the constraints of knowledge itself. We can only uncover the relations that exist, not those that do not exist, and most of the relations that we learn in our lifetimes are those edges that have already been discovered by others. Moreover, each field of knowledge reflects a different substrate and exists at a different stage of discovery, manifesting as a different network architecture that we might be able to learn. The constraint of knowledge's current architecture on our building practice becomes even more complicated when we consider the fact that knowledge itself is growing; our building potential thus changes each day as collective knowledge changes. This complexity motivates the use of models that couple two processes: the process whereby the network grows, and the process whereby the tissue or substrate beneath or surrounding the network grows. Modern models of vasculature are good examples. Recent studies have coupled growth models of the underlying biological tissue with growth models of the overlaid vessel network. In doing so, these models can explain the emergence of highly optimized transport networks in animal and leaf vasculature.[49] Such coupled models are strikingly appropriate for the practice of curiosity, where we envision that our walk on the knowledge network landscape depends not just on the contours of that landscape today but also on the patterns of wind, water, and other walkers that might change that landscape tomorrow. We grow curiously upon our changing earth.

*　　*　　*

Here we have taken the notion of curiosity as edgework and offered a concrete operationalization in network science. Along the way, we have proposed that the practice of curiosity is a practice of building knowledge networks. This proposition offers an interdisciplinary perspective on curiosity that is informed by neuroscience, psychology, philosophy, linguistics, and mathematics. Drawing on concepts and tools across these disciplines, we investigated the manner of network growth in our minds, and the potential to quantitatively characterize and model that growth. The view formalizes many of the intuitions that we have about the practice of curiosity, and by that formalization provides the foundations from which to construct explicit hypotheses that can be tested empirically in humans. We leave many open questions—some to be answered in the coming pages, and some to be left unanswered. Perhaps the most exigent, and the one we turn to in the next chapter, is the question of how we move in the conceptual space around us. How do our minds traverse those spaces as we build knowledge networks through acts of curiosity?

Curiosity may enable us to reach farther for ideas—to seek, search, and track with greater fervor and with greater dedication than would otherwise be our wont, leading us to stretch out tendrils into a broader knowledge network space. What is that reaching, stretching, and leaping? What are the kinesthetics of curiosity?

4 CURIOSITY'S GOT STYLE

What if risk traced a territory before it even accomplished an act?
—Anne Dufourmantelle, *In Praise of Risk* (2019)

From building blocks to Legos, from single words to opuses, what we build grows in complexity, but also in style. It develops according to a certain logic, whether the preferences of the builder, the qualities of the materials, or the constraints and encouragements of the environment. So too with our knowledge. Rooms of facts and whole buildings of expertise in a given field are curiously crafted according to the interests and needs of the individual as well as the vagaries of their personal histories. As that person changes over time, moreover, different rooms may fall into disuse and be repurposed (e.g., calculus for some people or iambic pentameter for others), while buildings might be razed to the ground and reimagined. This dynamic journey of individual curiosity depends, furthermore, on the era in which a person finds themselves, its economy and existing geography. Just as engineers turned from the steam engine to light-rail, from castles to skyscrapers (and tiny houses), so too do knowers change the sorts of structures they build over time. The knowledge architectures we build (and need) in a digital age are no doubt different from those required in hunter-gatherer societies or just before the printing press. And yet the idiosyncrasies of each curious person ultimately triumph in more ways than one might expect. Even at a similar juncture in history and with a shared commitment to harmony with nature, architects Antoni Gaudí and Frank

Lloyd Wright could not help but build in their divergently raucous and reposed ways, respectively.

So how do you build? If you were to pass through writer Norton Juster's iconic Phantom Tollbooth, would you gravitate toward Digitopolis or Dictionopolis? Do you prefer to use retweets and reposts as your raw material, or do you like hands-on projects? Or are you the sort to bury yourself in tomes and treatises, the dustier the better? Are you more likely to second-guess common sense or to curiously apply what you know to new cases? Or perhaps you find yourself perpetually on a quest to think ideas never thought before—notions with little relation to anything you have encountered along the way. Are you comfortable with not knowing things, such that the knowledge you build has perhaps more windows, doors, and skylights than your neighbor's, and maybe an infinity edge or two? Or do you like to be an expert of sorts, making it your hobby to know everything about Middle Earth or Hogwarts, a specific cuisine or restoration woodwork? Does your curiosity bend you to the page where, with brush or ballpoint, you experiment with worlds of your own building, the strange internal alchemy holding you spellbound? Or perhaps you take to the dance floor or athletic field, to the sea or high altitudes, to find in your body questions you never knew to ask and weathered wisdoms that answer. Do you like to chat about your curiosities—indeed, do you find it impossible *not* to chat about them? Or are you the sort to query in secret, just you and the air you breathe?

What, after all, is your style of curiosity? What are the design principles (and aesthetics) by which you build your knowledge networks? In this chapter, we investigate three archetypes of curious people, distilling them into three basic modes of practicing curiosity: the busybody, the hunter, and the dancer. We do so by canvasing the long history of Western thought, tracking the moments in which curious personages appear, and identifying their behaviors. The curious person is repeatedly described as 1) collecting stories, 2) hunting down secrets or discoveries, and 3) taking leaps of creative imagination. Sometimes the curious person is delighted to learn anything and everything, while at other times they become obsessed with a specific epistemic quest, and at still other times they explore whole new configurations of thought and world.

Whatever the era, each century seems to have its own version of these archetypes. In noting their distinctive behaviors, we attend to the unique kinesthetic signatures of each mode—that is, how the busybody, the hunter, and the dancer differently traverse conceptual and social space, and therefore weave differently valenced knowledge networks. While many people are predominantly one sort of curious person rather than the other, we close by considering Chicana feminist Gloria Anzaldúa, who aptly captures all three modes of curiosity in her one extraordinary writing life. On the whole, we set out to reimagine the ways in which styles of curiosity are, in many ways, styles of building our lives.

A NOTE ON MODELING

It is an odd thing to dive back into the history of thought not for the choice definitions of curiosity or the languorous theorizations of curiosity, or even the pockets of pontifications upon it, but rather for the people. Not for this or that now-famous personage who, having (supposedly) contributed something essential to the world's tale as we know it, is now assumed carte blanche to be curious, no. But rather for the literary figures who are deployed here and there, and the apparitions summoned forth, to embody curiosity. How is Curiosity said to live? Said to move? Said to talk? In what garb is Curiosity dressed? And what are their social habits, their fair-weather friends or stalwart companions? Here we dive into the history of the human soul to round up its curious characters. Attending to the use of the word *curiosity* itself, across the languages common to Western philosophy, we read over and over again about the busybody, the hunter, and the dancer, about collecting stories and telling tales, tracking down answers and chasing leads, leaping in the mind and dancing across the page.[1]

Prudently, we lay no claim to a final or exhaustive treatment of either these figures or their role in the history of thought. Nor do we assert that these are the only curious figurations in that archive. Instead, we aim to provide a provocative sketch of these models of curiosity. The busybody, the hunter, and the dancer each highlight a unique praxis of curiosity. Whether it involves collecting new bits of information, tracking down specific answers, or experimenting with breaks in tradition, each model illuminates a different modal dimension. They are not static representations of curiosity as such but rather dynamic depictions of how curiosity works, how it behaves, and what it does. They portray how curiosity moves. Throughout philosophical history, busybodying, hunting, and dancing capture specific kinesthetic signatures that map out different styles of knowledge network building in conceptual and social space.

More than literary figures, the busybody, hunter, and dancer are models in the technical sense, providing a vivid representation of certain features of curiosity.[2] Put simply, a model is a functional representation of a particular feature of the world; that is, it depicts what something is—or more precisely, what something is in terms of what something does. These three stylizations of curiosity are perhaps best characterized as idealized and phenomenological models.[3] That is, insofar as they produce an abstraction or caricature of curiosity, by isolating a set of properties or deliberately highlighting them, they are *idealized* models. And insofar as they draw out merely observable characteristics of curiosity, without postulating a theory behind them, they are *phenomenological* models. As such, they can be used to elucidate curiosity itself. Importantly, as models of curiosity, the busybody, the hunter, and the dancer are not mutually exclusive, nor do they necessarily represent the only forms of curious life. Nevertheless, their ambits are helpfully distinct and, to a certain extent, also complementary in the very architecture of curiosity.

STYLE #1: THE BUSYBODY

Perhaps the oldest, most consistent characterization of the curious person in the history of philosophy is as a busybody. At the most basic level, a busybody is busy making it their business to know anything and everything. They do not stop at their front door or immediate surroundings, at high school or what they need to know for work. They are constantly curious about all kinds of things, with no real rhyme or reason. As such, their curiosity is immensely generous. A bit of a butterfly, their wide-ranging curiosity results in a sprawling knowledge network, which lightly loops together an ever-expanding set of divergent experiences and disparate archives. Moreover, because the busybody is more excited about new fads and facts than old faces and familiar stories, their knowledge may reach far and wide, but it is rarely deep. For this reason, ancient thinkers suspected the busybody of being a superficial knower, a gossip and a jack-of-all-trades, rather than a self-reflective person capable of reining in their curiosity and focusing on the task at hand, whether family, or expertise, or the mysteries of modernity.[4] Whatever the historical social standing of the busybody's curiosity, however, their associated activities bear certain kinesthetic signatures. Here we ask, By what logic does the busybody behave? How does the busybody move? What are the geographical routes and social configurations of that movement? And what relation does that movement have to specific thought processes? Ultimately, how does this style of curiosity function?

In the ancient annals of curiosity, the busybody assumes an eager posture, replete with alert eyes, pricked up ears, and a quick tongue. First-century philosopher Philo Judaeus, for example, describes the body made busy by "curiosity." It is a body in which the senses—the eyes, ears, nose, tongue, and hands—"toss their heads, and frisk about, and rove about, at random, wherever they please." The busybody's ears, in particular, "eagerly receiv[e] every kind of voice, . . . never being satisfied but always thirsting for superfluity."[5] In contemporary terms, imagine someone who regularly keeps thirty tabs open on their web browser, or uses more pieces of tech than they have eyes or hands for. At the outset of the Middle Ages, Saint Augustine criticized this style of curiosity from a Christian perspective. Lamenting its spiritual perils, due to its fleeting interest in fleeting things, Augustine confessed to taking illicit pleasure in gossip, theater, astrology, astronomy, sports, and even nature watching.[6] Philosopher and mathematician Blaise Pascal replicates this critical sentiment in the modern period when he insists that curiosity is nothing but an expression of idleness. "We usually only want to know something," he writes, "so that we can talk about it."[7] If curiosity merely collects fodder for our inner chatterbox, it seems essentially vain. From our current vantage point, however, it is possible to reconceptualize this eager busybody as exuberantly attuned to the wide wildness of the world and hence able to make quick work of the edges around them.

Of the many historical depictions of the curious busybody, perhaps the two most memorable are offered by Plutarch, a first-century Roman essayist, and Martin Heidegger, a twentieth-century German philosopher. In his essay "On Being a Busybody," Plutarch characterizes the busybody as a *polypragmon*—that is, someone involved in many affairs (*poly-pragmatos*). Given their ever-expanding interests, you can find the busybody at the seaport, in the town square, at the theater, or in the public assembly, fluttering from place to place collecting news and swapping tales. In a vibrant passage, Plutarch describes them thus:

> The busybody, shunning the country as something stale and uninteresting and undramatic, pushes into the bazaar and the marketplace and the harbors: "Is there any news?" . . . If, however, someone really does have something of that nature to tell him, he dismounts from his horse, grasps his informant's hand, kisses him, and stands there listening.

The busybody is wildly curious about everyone else's business and is committed to ferreting it out. They are clearly not at their own home, attending to their own business, or focusing on their own self-development, and they would rather die than live among rolling hills. Wherever information is heatedly generated and exchanged, that is where the busybody wants to be. From Plutarch's perspective, this kind of curiosity unfortunately devolves into a "disease," where the "passion for finding out whatever is hidden" becomes fueled by "a desire to learn the troubles of others."[8] Given this danger, Plutarch recommends curiosity be renounced altogether, but those incapable of such severity can fruitfully redirect it toward history, biology, or astronomy.

Heidegger reprises Plutarch and Augustine's critical perspective many centuries later. In *Being and Time*, he argues that curiosity marks the modern individual who, consumed by the advances of a technological age, is increasingly incapable of existential reflection and authenticity. This individual is curious insofar as they want to know something quickly or experience something briefly, but then move on to new intrigues. They are consumed by passing interests. For Heidegger, curiosity is a "care to see not in order to understand what it sees, . . . but only in order to see." Like those before him, Heidegger also links curiosity to idle chatter, which "does not communicate" in some deep and meaningful way but rather speaks "by gossiping and passing the word along." By capitulating to curiosity and idle talk, the modern individual only ever skims the surface. Over time, this creates a zombielike existence, marked by "groundless floating," in which they never really see or say anything at all—even if they never really stop seeing and saying things.[9] For this reason, Heidegger insists, a new attunement to the world is needed: one of care rather than curiosity.

Despite the historical disparagements it has sustained, the busybody style of curiosity is deeply compelling (and deeply human). At its best, busybodying signals an openness to experience, irrespective of its utility and even celebrative of its excesses.[10] It can function as an act of radical listening—one that enrolls the curious person in the project of dialogical existence, living through—rather than on or without—all the other curious beings and things in the world.[11] And it is more likely to stretch beyond traditions, canons, and dominant knowledge systems to hear what often goes missing or falls through the cracks of prevalent social logics.[12] From a network perspective, the busybody's conceptual space is characterized by quick associations, disparate strands of information, and loose knowledge webs. They are interested in conceptual stimulation: whatever lies outside their current knowledge grids. Being easily pulled in different directions, they pursue their interests quickly, covering short distances and shallow depths. They then use the new threads they collect to extend their knowledge grids in a random, delightfully disordered way, rather than build organized systems. This, at least, is the busybody's kinesthetic signature, the way it moves curiously within social and conceptual space.

STYLE #2: THE HUNTER

In contrast to the erratic energy of the busybody, the hunter's curiosity is focused, sustained often across arduous investigations and solitary quests.

The hunter explores, traces, and tracks with a singularity of purpose.[13] They investigate the animal (or mineral) along with its habits and habitat. They track it, with great patience and persistence. They spy it from afar and refuse to let it out of their sight. They instinctively become the thing they follow, imbibing its spirit, its intelligence, and its secrets. Similarly, curious people hunt for answers, they explore new terrains, and they track down hunches or leads in search of a new discovery. With well-honed interests, they are at work in the wee hours, typically lost in thought, and difficult to extract from the library or archives, the lab or forest, the café or community space. The strength of the current driving their curiosity is undeniable, almost unbreakable. What churns behind their eyes is a lodestone of desire, not so much to solve as to understand, not so much to acquire as to be submerged within a landscape and therein find a new path forward. In any case, they are bent—bent over a task or under a question. And it is in this sense that they listen. Hunters attune themselves to what they hunt. This is a targeted mode of curiosity that, in singling something out and pursuing it, invokes a whole web of curious practices. What are the structural and kinesthetic dimensions of such a curiosity? What is the correlation between this sort of curiosity and its specific thought processes or patterns?

Plutarch, who conceptualized curiosity primarily in the busybody mode, nevertheless grants that curiosity can sometimes surface in the hunting mode. In the same essay mentioned above, Plutarch discusses the rare occasion when the busybody's curiosity is redirected from the marketplace and seaport to the academy, where the person might begin to study astronomy, biology, and history. In such a situation, their curiosity is no longer distracted by each new interest, but becomes solemn and focused. He writes,

> And as hunters do not allow young hounds to turn aside and follow every scent, but pull them up and check them with the leash, keeping their sense of smell pure and untainted for their proper task in order that it may keep more keenly to the trail, . . . so one should be careful to do away with or divert to useful ends the sallies and wanderings of the busybody.[14]

The hunt for knowledge demands incredible dedication. The hunter must purify themselves, refrain from all manner of diversions, and pursue only their one true quest, their signal question. This pursuit requires a singularity of purpose and a narrow selection of what, in fact, one is driven to explore, what relationships one hopes to uncover, and what connections one wants to make. While Plutarch—much like Augustine several centuries later—will insist that even the hunter's curiosity is not proper to self-reflection or spiritual devotion, nevertheless its disciplined attention certainly commends it.[15]

At the turn of the Renaissance, this hunting curiosity—with its keen mentality and earthly interests—became slowly consistent with reflective existence. Poet and author Geoffrey Chaucer's *Canterbury Tales* dramatizes the shift in the character of the Monk, who travels widely, gabs garrulously, and bears an unusual obsession with hunting wild hares. In so doing, the Monk disregards the traditional spiritual practices of withdrawal, silence, and wondrous lingering. Of a similar ilk, Bishop Richard de Bury, a fabled fourteenth-century book collector, became renowned for his love of books. His obsessive interest was so keen, his efforts to build a collection so feverish, and his curiosity so insatiable that he was accused of heresy and threatened with excommunication.[16] In his treatise *Philobiblon*, he not only defends himself but enjoins others, too, to taste the delights of chasing after books. He speaks glowingly of his habits "of hunting [*venandi*] as it were some of the most delightful covers" across public and private libraries, in Athens and Rome, Britain and Bologna, but most of all Paris.[17] Driven to books chiefly in search of wisdom, whether of secular sages or religious seers, de Bury represents a growing Renaissance acceptance of absorptive curiosity.

By the modern period, curiosity's hunting mode was relatively well regarded. Philosophers Friedrich Nietzsche and Jacques Derrida are herein exemplary insofar as they commend a certain avid interrogation of history and language in order to flush out the buried potentialities of humanity. Writing in the nineteenth century, Nietzsche endorses the hunting mode as necessary for philosophical inquiry. Such a hunt ferrets out the contours of history so as to patiently track within them humanity's worst flaws and greatest potentials. He writes that of all the things that "excite his curiosity, . . . the whole history of the soul so far and its as yet unexhausted possibilities—that is the predestined hunting ground for" someone who, like him, is a "lover of the 'great hunt.'"[18] Somewhere, deep inside these historical vicissitudes lies not only the reason for why we do what we do but also the road map for how we might do otherwise. Just as Nietzsche commends a curious hunt into history, so more recently Derrida endorses a curious hunt into language. There, in the ever more organized catalogs of meaning, lie still other dormant possibilities—words that might be loosed from their confines and set free to fly. The curious hunt through the thickets of language—"to track, to sniff, to trail, and to follow," even if "just to see" where each might lead—is perhaps the most effective way to unsettle conceptual binaries and semantic borders. In that hunt, Derrida insists, there is as much a "following" as there is a "ferreting out," such that the pursuit is also, fundamentally, a practice of listening.[19] It is through a compulsive listening and a dogged trailing that philosophical insights into human capabilities and linguistic capacities are to be found.

Due to the hunter's focused and dogged commitment—their capacity to fall down rabbit holes, get lost in the tunnels, and only resurface years later—they build knowledge networks that are architecturally quite different from those of the busybody. As the hunter hunts, they build targeted connections and craft tight knowledge networks, fastidiously shaped and deeply researched. Refraining from the haphazard temptations of novelties and secrets, they undertake a pointed pursuit of information within a predetermined, even if lightly mapped, subject area. Moving swiftly and slowly by turns, they steadily mold discrete extensions of existing knowledge networks in such a way that both individual and collective knowledge grows. In supporting the progress of knowledge overall, however, the hunter also remains open to discovering an outlying fact that might necessitate the rupture of current knowledge schemas and thereby lead to radical innovation. Whatever their relation to social structures, the hunter is fundamentally relational insofar as their footpath is determined in advance by that to which they listen. Indeed, curious hunting requires a certain humility, mobilized by following and carefully attending to what they track. As such, their knowledge building involves the delightful coconstitution of (un)knowers and (un)knowns.

STYLE #3: THE DANCER

Against the gathered threads of the busybody and the projectile path of the hunter, this third mode of curiosity thrives on breaks, ruptures, disruptions, and a staccato rhythm. The dancer leaps. They lift their feet from the ground and are caught up in a series of arcs and arches, brief forays into the sky. Across the ages, curiosity has consistently been aligned with a practice of uprooting. One might think, for instance, of the vagabonds and travelers (*curiosi*) of the Middle Ages, or Heidegger's description of the curious person as "never-dwelling-anywhere [*Aufenthaltslosigkeit*]."[20] This uprootedness, however, is not always a steady state; sometimes it is the result of a choice or the inspiration of a moment. It may be the leap of creative imagination, when you suddenly take to pen and paper, or acrylic and mylar, or body and beat, to explore with serendipitous inklings and exuberant hopes the limits of the seeable and the sayable. Dancers query on the wings of inspiration, nonplussed by the lack of guardrails or precedents, and ready to take risks. So what is the dancer's kinesthetic signature? How does the leap correspond to the formation of curious thoughts? And in what ways does the dancer build? By what architectures are they buoyed?

Let us return to Nietzsche. For him, the ideal reader, thinker, and social critic is a person of "jubilant curiosity [*Neugier*]."[21] That person is adventurous and courageous, drawn to risk and to discovery.[22] That person, in fact, loves to dance. For Nietzsche, dance represents a kind of irreverent, embodied exploration of existence, and he opposes it to the sad, somber practice of transcending the senses—something, he argued, only boring philosophers, or mummies as he called them, attempt.[23] As much as he would "believe in a god who could dance," he *does* unequivocally believe that minds, too, can dance.[24] What is a thinking that can dance? What does it look like? What does it sound like? How does it move? And how is it moved? He writes that one must be "able to dance with the feet, with concepts, with words . . . with the pen."[25] For Nietzsche, it is easy to tell a book written by a dancing mind from another written by a more plodding character. Books written by dancers can themselves dance. "It is [my] habit," he writes, "to think outdoors—walking, leaping, climbing, dancing. . . . [My] first questions about the value of a book, of a human being, or a musical composition are: Can they walk? Even more, can they dance?"[26] It is this dance that, for him, marks the mind of a jubilantly curious social critic.

While this dance itself has several different characteristics, depending on whose description one consults, perhaps its most consistent expression, its signature move, so to speak, is the leap. As contemporary theorist Alan Watts sees it, curiosity will always involve leaps. "By replacing fear of the unknown with curiosity," he muses, "we open ourselves up to an infinite stream of possibility . . . pushing our boundaries [and] leaping out of our comfort zones."[27] Michel Foucault underscores this energetic propulsion, not in a psychodynamic context but in a more traditional academic sense. He disregards philosophy's long-standing dismissal of curiosity across the ancient and medieval periods. Instead, he endorses it wholesale as precisely what allows one "to throw off familiar ways of thought." Well within the Nietzschean tradition, Foucault understands curiosity to be the foundation of all good philosophical thinking. He describes this curiosity-driven thought as one that "bring[s] an oeuvre, a book, a sentence, an idea to life." Waxing poetic, he continues that "it would light fires, watch the grass grow, listen to the wind, and catch the sea foam in the breeze and scatter it." Such curious thought, moreover, is consistently marked by "scintillating leaps of the imagination."[28] For Foucault, these leaps form part of a larger choreography of the curious mind.

In one of the most powerful passages in the history of philosophy, Nietzsche captures the spirit of these leaps, alongside the counterweight of an incurious "lumbering":

> Philosophy leaps ahead on tiny toeholds; hope and intuition lend wings to its feet. Calculating reason lumbers heavily behind, looking for better footholds, for reason too wants to reach that alluring goal which its divine comrade has long since reached. It is like seeing two mountain climbers standing before a wild mountain stream that is tossing boulders along its course: one of them light-footedly leaps over it, using the rocks to cross, even though behind and

beneath him they hurtle into the depths. The other stands helpless; he must first build himself a fundament which will carry his heavy cautious steps. Occasionally this is not possible, and then there exists no one who can help him across. What then is it that brings philosophical thinking so quickly to its goal? Is it different from the thinking that calculates and measures, only by virtue of the greater rapidity with which it transcends all spaces? No, its feet are propelled by an alien, illogical power—the power of creative imagination.[29]

The dancing mind of the curious person moves by leaps of creative imagination. It is these leaps, apparently, that constitute good philosophy, better books, and the best of minds.

If the busybody arranges bits of information into loose knowledge webs, while the hunter organizes specific tracts of information into tight networks, the dancer may rupture those networks by either jumping to a new idea or throwing existing ideas into a new frame.[30] Driven neither by secrets nor necessity, the dancer is an experimenter, breaking with traditional pathways of investigation. Their conceptual space is therefore characterized by discontinuity, by the creation of new concepts, and by radically remodeling what is and can be known. Importantly, the dancer dances anywhere and everywhere, with or without companionship, erratically practicing an artistic irreverence for existing social and epistemic schemas. While the dancer's leaps are temporally brief, much like the busybody's forays, they cover a dramatic distance across the landscape of conceptual geography. This is why those leaps of imagination, those bursts of innovation, can put everything—from concepts to customs—into radical question. While harnessing the potential of creativity, then, dancing also courts the hazards of unmoored risk and social disobedience.[31] That very act of imagination that leaps into a new idea or a new epistemic space is the same act of imagination that gambles with what already exists.[32]

THREE STYLES, ONE LIFE

Busybody, hunter, dancer. Three models of curious behavior, three sets of curious practice. Without positing the nature of curiosity, these three figures draw out its observable kinesthetic signatures, thereby serving as phenomenological models. And without capturing all the intricacies of curiosity in action, they nevertheless isolate distinct habits of curious movement, speed, and depth along knowledge grids and social networks, thereby serving as idealized models. Noisy though they may be, these models are accountable to and illuminative of several basic forms of lived curiosity. As with fashion in other forms, furthermore, some people may gravitate toward one style more than another, while others might don each in their turn, depending on the context, the project, or the mere beckoning of delight. Each style is there for anyone and everyone, and yet each individual person will navigate them differently. And there are no doubt other styles yet to be culled and characterized. If the busybody is like a butterfly and the hunter is like a hound, what style of curiosity might an elephant or a prairie vole, an octopus or a crawfish, an eagle or a mosquito best communicate? How might the hummingbird coalesce multiple styles in its dazzling animacy, with its swiftness and torpor, a fierce focus and a bit of flitting. Indeed, by what bestiary might the delights of inquiry be further illuminated? Whatever the ultimate pantheon of creaturely curiosity, we turn in this final section to explore the confluence of all three styles in one life: the life of Gloria Anzaldúa.

Anzaldúa was a Chicana author, poet, and activist in the late twentieth century whose books *This Bridge Called My Back* and *Borderlands / La Frontera* firmly established her as a third-wave feminist icon. Her work addresses critical issues of feminism, intersectionality, and coalition, alongside the transformative power of writing and the necessity of spiritual liberation. Anzaldúa changed the face of feminism, insisting that it be shattered by the experiences and knowledges of queer women of color and then reconstituted along a different track, on a new terrain. Anzaldúa's books, selling hundreds of thousands of copies and going into multiple editions (and translations), are now cornerstone texts in American studies, ethnic studies, LGBTQ studies, and feminist theory. Her poetry, fiction, and prose continue to inspire whole generations of writers, whether among the ranks of academicians or activists. Growing up as a seventh-generation American and migrant worker in Texas, Anzaldúa was keenly drawn to the page. Although "criticized for being too curious," she recalls being "hungry for more and more words," and as an adult having "an incredible hunger to experience the world."[33] It is that hunger, in its multiple modes across time and especially in her writing life, that we aim to elucidate here.

Anzaldúa's curiosity ran rampant, unrestrained by societal expectation or academic discipline. She recalls attending grade school in the Edinburg Independent School District on the north side of McAllen, Texas, just twenty miles from the Rio Grande. She would go to the library to escape being Chicana, being a girl, and being in pain with a severe endocrine condition. She would read and read and read. "I'd read everything in the library. Everything: encyclopedias, dictionaries, Aesop's Fables, philosophy—I started reading all these heavy books. I literally went through all the shelves book by book."[34] In adulthood, the same instinct predominated when she was procrastinating, but also when she was feeding the muse (if the two are even separable). She would sponge read and binge read by turns, traversing Spanish and Mexican literature, white feminism, lesbian feminism, Indigenous mythology, shamanism, and philosophy. She had a special penchant for chaos theory, Lucha Corpi mysteries, and Laurell Hamilton vampire novels (she made no apology), alongside philosophers Nietzsche and Arthur Schopenhauer, Jeffner Allen and María Lugones. "Reading voraciously in all disciplines," she would collect and gather, building a vast web of meaning irrespective of its ultimate usefulness for her life or relevance to her writing projects.[35]

Anzaldúa's curiosity was also dogged and tenacious, refusing to let up or give in. This is true of her search for self-identity and coalition, but it is most true of her writing. After all, for Anzaldúa, writing was the primary path to knowledge and therefore the vessel of curiosity, which for her was a mix of questioning and creativity. Often turning her unusually rich source materials into composition fodder, she was an exacting craftsperson. Notoriously scrupulous about the writing process, it was not uncommon for Anzaldúa to stretch pieces over a dozen or more drafts, taking months, years, or even decades to bring things into being. In writing, she would revise over and over again. Perhaps something of a perfectionist, she was deeply attuned to getting it right—the right word to link to other words, the right symbol or image, the right story at the right moment. As a writer, she would say, you have to "suck on the problem" until you can "capture" the images in your "net of words. . . . You rework the material until alignment and balance occur. Then you try to get each section to lock into sync with every other section, lock and pulse together the undercurrent rhythm of the whole piece."[36] In this exacting effort to build tight, purposive, and thereby effective networks, Anzaldúa kept her eye on the prize.

For Anzaldúa, the process of coming to know is not all sponge and capture. It is also, and perhaps most fundamentally, a process of creativity. To write, she insists, one must sink into *nepantla*—the Nahuatl term for a liminal or borderland state, where concepts dance and links leap. Border dwellers, or *nepantleras*, "position themselves in the crack between worlds," in the fuzzy edges of sense-making frameworks, where less is set and more is supple. Here in the in-between space, in the space of bridging and breaking, nepantleras can "construct alternative roads," creating whole "new topographies and geographies" in their wake. Anzaldúa herself was a nepantlera through and through. In her writing, she allowed things to decompose before her eyes, and then set to work "sifting" and "sorting" things through, "[re]arranging" and "patching" things back together, so as to craft a new story. The process always began with her willingness to be worked over by *remolinos* or whirlwinds, vortexes of new ideas and social change that set existing knowledge adrift, sending nodes and edges helter-skelter. She would then call on "the 'connectionist' or web-making faculty, one of less structured thoughts, less rigid categorizations, and thinner boundaries." Activating this faculty, she was able to take "leap" after leap in the hinterlands, until slowly but surely her writing would "start to dance."[37]

Anzaldúa wrapped all three modes of curiosity into one. Weaving together her urge to collect widely, to track scrupulously, and to leap imaginatively, she would move in and out of each mode on different registers, within the space of hours, days, or even weeks. Engaging the modes together allowed her to generate the wide-ranging, raw, and yet piercing pieces that make her so memorable today. Anzaldúa's goal was to cultivate a way of knowing that was not riven by unbearable (and ultimately unsustainable) divisions; rather, she aimed to create within the in-between space. How then, she would ask, can we become skeptical of conventional knowledge structures and instead nurture other ways of knowing? How can we build epistemic architectures within the borderlands so that the interconnectivity of things is honored rather than erased? How can art infuse thought? A deep fan of chaos theory and quantum physics, she had a sense that shuttling assiduously between loose filaments and liminal leaps gave her the best chance of developing frameworks flexible enough to honor the complexity of the world. This too is curiosity.

* * *

When we reconceptualize curiosity as a growth principle, we can ask how knowledge grows. Does it weave around preexisting trellises, spread invasively into nearby beds and embankments, or does it quietly coalesce year after year into tree rings, ever so patient and compact? Does it cycle through seasons of inflorescence and hibernation, within a larger ecosystem of dynamic growth and decay? Pivoting toward practice, this paradigm permits us to ask not whether this or that person is curious but rather how they are curious, and at what times and in what ways. In what manner do they seek to know, and by what style is that knowledge knit or mishmashed together? By what alchemy are new nodes and edges built, and then razed and rebuilt? Curiosity here becomes a series of modes or a set of practices. We have begun modeling those modes through philosophical literature. By illuminating the literary figures of the busybody, the hunter, and the dancer, we can more productively identify and explore people's different styles of curiosity and their respective kinesthetic signatures. Perhaps most fundamentally, we can notice our own patterns of knowledge building and lean into our strengths. Alternatively, we can experiment with building our knowledge differently, trying out different strategies and tactics, and reevaluating what we are building *for*. After all, the style of our curiosity deeply informs the style of our lives.

In retrospect, we often wonder if these styles have still deeper cosmic roots. What if they were traceable not only in historical thinkers and texts but in the set of possible dimensions of space too? One could imagine that the human being, in zero dimensions, would exist as a point, never moving, never seeking, never really curious. In one dimension, however, that same human would walk in a line, tracking much like the hunter. Then in two dimensions, they would reach out to everyone in a hub-and-spoke architecture (à la the busybody). And in three dimensions, they would track or collect in two dimensions, but then leap up in the air, by a third dimension, to another space (à la the dancer). Finally, adding the fourth dimension of time, our imaginary human could weave among curious styles, spending moments, periods, or eras as a point, a line, a plane, and a volume, stringing together archetypes as a pattern of colored beads on the thread of their lives. Perhaps, indeed then, these three curiosity archetypes can be thought of as the most basic sorts of seekers. But can these historical and seemingly a priori modes be traced out empirically, both within human behavior and deep in neural network architecture? And if so, how might that connection inform communication theory and pedagogy going forward? It is to these intriguing questions that we turn in the next chapter.

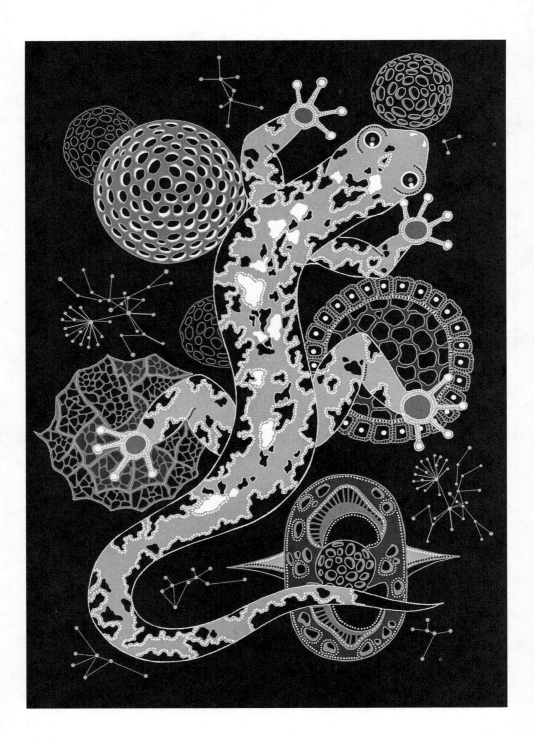

5 WEBS OF KNOWLEDGE

> I am struck with the truth, that far more of our deepest thoughts and feelings pass to us through perplexed combinations of concrete objects, pass to us as involutes (if I may coin that word) in compound experiences incapable of being disentangled, than ever reach us directly and in their own abstract shapes. . . . Man is doubtless one by some subtle nexus, some system of links, that we cannot perceive.
> —Thomas de Quincey, *Autobiographic Sketches* (1853)

What is the shape of knowledge? If knowledge were a building, would its architecture be angular, rectangular, squarish, and concretish as the 1960s' Churchill College, Cambridge, on Stories Way? Or would it be curvilinear, along the exterior and interior, as the unfinished Basílica de la Sagrada Família in the Eixample district of Barcelona? If knowledge were a rock, would it be porous, soft and crumbling before us? Or would it be igneous, smooth and glassy to the touch? If knowledge were a kind of matter, would it be composed of atoms floating in an ether or elements held rigidly in a crystalline scaffold? If knowledge were formed by the fluid of experience, would it be the distended contours of a riverbank after a summer's rain, the mold of a pool in the constraining crook of a tree branch, or the dome of a water droplet on a spring leaf, bending in obeisance to the physics of surface tension?

These questions are slippery. (Like the belly of a red-spotted salamander, running through my young fingers before diving under the nearest rock in the forests of Pennsylvania.) How do we gain traction on them?

Well, we have to ask ourselves, What do we know about knowledge? We know that knowledge both does and does not exist. We hand each other knowns, and we face unknowns. Knowledge is apparent when set in relief against not-knowledge. Sometimes what we do not know exists close to what we do know; think of the taste of ripe California avocados on coffee Heath bar crunch ice cream. While we do not know the taste of the combination, we know the taste of the two components. Sometimes what we do not know exists far away from anything we know now or have ever known in our lifetime; think of the feeling of jumping on the moon while staring out at the earth. Of none of these components do we have any prior knowledge. From close to far, the distances between bits of knowledge are of universal proportions.

But beyond and within those proportions lies an architecture, a scaffold, a shape. A shape upon which we move. Whether we move through knowledge in a line (as a hunter), a plane (as a busybody), or a volume (as a dancer), we move in and through a shape. What is that shape? The shape of knowledge is much like that of Swiss cheese . . . in space. Like Swiss cheese because distributed throughout our knowledge (the flesh of the cheese) are pockets of unknowns (the holes in the cheese). For example, I know that when seeds sprout, the first leaves that open to the sun to receive energy are the cotyledons, and then the second leaves to open are the foliage leaves.[1] I also know that cotyledons are internally homogeneous and their peripheral edges are smooth. In contrast, foliage leaves can have rough boundaries, intricate geometries, and complex vasculature evident to the naked eye. I can think of all of these facts as points of cheese along the perimeter of the hole. Inside the hole between these points is the question, Why do plants need two sorts of leaves that differ so drastically in their architecture given that both types effectively gather the energy needed for development? Surely an answer exists to this question, but it is not housed in my own mind and therefore that portion of my knowledge is simply an empty space. The same is true of others. Any human's knowledge can have empty spaces, reflecting little sectors of personal ignorance. And in fact, any *society's* knowledge can have empty spaces, reflecting sectors of sociocultural ignorance arising from a lack of funding for relevant research, and a devaluation of the areas of inquiry pursued by thinkers of marginalized genders, races, and ethnicities.[2] So knowledge is like Swiss cheese, a flesh of knowns interwoven with pockets of unknowns. Despite the caloric content, that analogy is not fully satisfying because not all of my areas of ignorance are circumscribed by areas of expertise. Hence knowledge must be like Swiss cheese *in space*, wafting, pausing, drifting in the blank darkness. A single human's knowledge is much smaller than collective human knowledge, the body of knowledge formed when we pull together the individual knowledge of every human both alive today and existing in the past (if that were even possible). In fact, even collective human knowledge is much smaller than the knowledge that could be acquired by the human species as time goes to infinity. Our knowledge is indeed so small that while jumping on the moon, we might not even notice the bit of Swiss cheese floating through the void against the backdrop of the sun-tinted earth.

KNOWLEDGE AS CONTINUOUS VERSUS DISCRETE

To the human eye, cheese is voluminous, solid, and continuous rather than discrete. Is the same true of knowledge? Let us consider the alternate hypothesis that knowledge is discrete or consisting of individual parts that are detached from others. Is knowledge composed of distinct bits separated by a noninfinitesimal distance? Answering that question will require that we examine how knowledge—or at the very least a sector of knowledge—is composed.

Consider the work of Maximilian Fürbringer, a nineteenth-century German anatomist, who constructed an evolutionary tree of all the extant and extinct bird groups using fifty-one anatomical and morphological characteristics, prior to the advent of molecular techniques.[3] The tree is a clear visual presentation of the fact that groups of birds exist, and there are marked differences between the groups. We do not find a continuum of birds with infinitesimal differences like the continuum of colors on the visual spectrum from burnished red to dusky violet. Instead we find distinct birds, individual birds, discrete birds. Now consider the periodic table of elements, which is a clear visual presentation of the fact that distinct elements exist.[4] Each element differs from all others not by an infinitesimal change in charge or mass but rather by fixed units of protons. Now consider the table of elementary particles in the standard model of physics, from which we appreciate the distinctions between quarks, leptons, gauge bosons, and scalar bosons.[5] Again, the particles do not mark infinitesimally separated points on a line but instead differ from one another in a quantized manner. No sheer chiffon continuum will ever appropriately drape the shape of this knowledge.

From these examples, we can conclude that matter—from the very small to the very large—is discrete rather than continuous. But we might push further to ask whether the discrete nature of matter or the physical world tells us anything about the nature of thought or the world of the mind and soul. The latter is arguably more difficult to see and measure. Yet we have some evidence from simple contemplation; recall polymath Benjamin Franklin as a twenty-year-old in 1726 demarcating the thirteen virtues that he wished his life to exude: temperance, silence, order, resolution, frugality, industry, sincerity, justice, moderation, cleanliness, tranquility, chastity, and humility.[6] Discretization is evident even in the modern era, where we might examine the human behaviors or skills that stem from the mind. Our culture defines skills as discrete abilities, which can be listed, enumerated, and bulleted, whether on an advertisement or a curriculum vitae. Complementing Curiosity's personal ad reprinted in chapter 2, Curiosity's *job ad* might read as follows: the ideal candidate for our position would have demonstrated skills in open-mindedness, inquisitiveness, and voraciousness. Different sets of skills are needed for different sorts of jobs; some are taught through higher education while others are innate or acquired from social experience.[7] In both cases—virtues and skills—it is difficult to define any true continuum along which the items in the list exist.

We can draw perhaps even more convincing evidence for the discretization of knowledge from the utility of logic. A systematic study of the valid rules of inference, logic is in part built upon the notion that ideas can be discretized and manipulated as symbols. Recall philosopher Gottfried Wilhelm Leibniz's claim that an alphabet of human thoughts must be invented, replete with a fixed set of letters that can be combined to

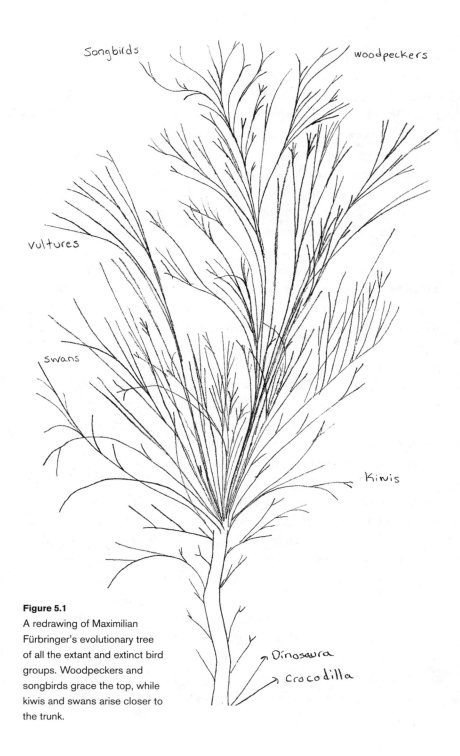

Songbirds

woodpeckers

vultures

Swans

Kiwis

Dinosaura

Crocodilla

Figure 5.1
A redrawing of Maximilian
Fürbringer's evolutionary tree
of all the extant and extinct bird
groups. Woodpeckers and
songbirds grace the top, while
kiwis and swans arise closer to
the trunk.

make the words by which everything can be discovered.[8] Our thoughts can appear as discrete bits, just like the letters of the alphabet, and just like the words of a language. Yes, language too is discrete, not just in its representation of concepts, but also in its practical reflection in symbols. If we had just the perfect number of letters and if words in the English language were approximating a continuum, we would guess that the number of words in the language would be a^n, where a is the number of letters in the alphabet (26) and n is the number of letters in a word. Because the average English word is 4.7 characters (although the longest common word is 45 characters), we would expect approximately 4,470,691 words. In contrast to our expectation, the second edition of the twenty-volume *Oxford English Dictionary* contains 171,476 current words and 47,156 obsolete words for a total of 218,632, or 5 percent of the number we predicted. Our language appears to be sparser than expected given the size of our alphabet, raising the question of whether it could ever be a continuous symbolic representation of concepts.

Of course, it is possible that our vocabulary is smaller than expected given our alphabet size because distinct roles are played by vowels and consonants, or simply because our minds have limited memory and encoding space, and find it easier to use and manipulate distinguishable elements. To probe the mechanisms of those limits, we must turn from the mind to the brain, where real estate is limited, and where cognitive processes of memory and encoding are constrained by the biophysical realization of computation. Here we find that even the wetware of neural circuitry works with discrete elements (neurons connected by synapses) according to a discrete computational code of cell activation (or so-called *spikes*). Might each spike be a letter in the brain's alphabet? To answer this question, neuroscientists have devised experiments to catalog the responses of neurons as they experience the world. Such experiments reveal a vocabulary in which each neuron's spike is a letter in the alphabet, and the set of neurons that spike at a given time is a word, and the set of words produced by the whole neural system as it experiences the world is the vocabulary.[9] The language is replete with an internal structure (including a thesaurus!) reminiscent of the design of engineered codes. The discrete nature of neural circuitry, neural codes, and human knowledge suggests the possibility of a structural similarity—or isomorphism—between the modeler (brain) and the modeled (world). Humans may build discrete knowledge of the world, in part supported by the fact that each bit of knowledge can be encoded in a discrete "word" of neural activity.

KNOWLEDGE AS A NETWORK

Hence at some level of description, knowledge can be thought of as discrete rather than continuous. As we experience it, knowledge reflects a discrete world, symbolized in a discrete language and processed according to the vagaries of our discrete brains. Our good friend the Swiss Cheese, wholly full of integrity, deferentially steps aside to make way for a new model. And what might that model be? Should we choose a collection of cheese bits (crumbs, flecks, proteins, and molecules), like a bag of coins or compilation of beads? A periodic table of elemental information? A standard model of particulate understanding? An agglomeration of atomic facts? Better, but not as far as we need to go. Any model in which knowledge is treated as isolated grains of sand misses a key fact: that bits of knowledge are connected to one another by relations. Note the following passage in mathematician Henri Poincaré's *La Science et l'hypothèse*: "The aim of science is not things themselves, as the dogmatists in their simplicity imagine, but the relations among things; outside these relations there is no reality knowable."[10] Or similarly a century earlier in naturalist Lorenz Oken's *Lehrbuch der Naturphilosophie*: "Science is a series of necessarily inter-dependent and consecutive propositions."[11] These two scientist-philosopher hybrids state frankly what we intuitively appreciate: we cannot truly understand any item of knowledge until we understand its relations to other items of knowledge.

Given the discrete and yet interconnected nature of knowledge, let us then consider the utility of a network model. Here we might represent each bit of knowledge as a node in the network, and we might represent a relation between two bits of knowledge as an edge or a link between them. Such a model is conceptually evident in John Dewey's *Democracy and Education*, where he writes,

In other words, knowledge is a perception of those connections of an object which determine its applicability in a given situation. Thus, we get at a new event indirectly instead of immediately—by invention, ingenuity, resourcefulness. An ideally perfect knowledge would represent such a network of interconnections that any past experience would offer a point of advantage from which to get at the problem presented in a new experience.[12]

What Dewey intuited in concept can now be formalized in mathematics. The resulting network model can be used to describe and understand many different sectors of knowledge, from common sense to emotional intelligence to small talk. Network models can also be used to understand formal areas of scholarly inquiry, including philosophy, history, and science. Whatever the sector of knowledge, concepts and relations live in (and flow among) minds, oral traditions, articles, books, patents, blogs, magazines, newspapers, practices, tools, and software.

Knowledge networks are constantly evolving. With each discovery, we add a new node (an idea, a model, a theory, or a bit of evidence) and connect it with an edge (of relation, of dependence, or of support) to the rest of the network. In distinct academic disciplines, knowledge networks have been formalized differently, by a specific set of thinkers and historically developed systems of thought. These specific origins coupled with the nature of the subject of inquiry itself can together lead to different network architectures characteristic of each field. A helpful mental picture can be found in philosopher, physician, and marine biologist Ernst Haeckel's exquisite drawings of radiolaria (sea protozoa; nearly 150 new species of which he named), poriferans (sea sponges), and annelids (segmented worms) in the 1870s.[13] The radiolaria, in particular, evince incredibly intricate network architectures in their 0.1–0.2 millimeter elaborate mineral skeletons, typically composed of silica. The spindly spicula of *Sphaerozoum variabile* may remind one of the lengthy tails of history, tracking the linear progression of lives through a series of events.[14] The regular mesh of *Orosphaera huxleyii* may remind one of the highly structured dependencies between mathematical notions in linear algebra.[15] And the cavernous loops of *Octopyle stenozona* may remind one of the structure of early vocabulary in toddlers as they learn to speak about their world in a way that illustrates the existence of many small holes in their understanding.[16] The diverse network structures of different areas of knowledge mirror the diverse network structures of our world.

WALKING ON KNOWLEDGE NETWORKS

Beyond the formal areas of collective knowledge, it is of interest to consider the knowledge network inside each of our heads. That individual network is constantly evolving and growing in manners and directions that are idiosyncratic to each of us. Curiosity flourishes at the interstices of the knowns and unknowns, at the boundary between the actuality and potentiality of knowledge networks. We walk along the links that connect bits of information, and then stand at the void and shout across. With poet Walt Whitman, we "think a thought of the clef of the universe, and of the future."[17] When we find or discover knowledge that fully or partially fills the void, we have built a new bit of scaffold in our knowledge network.

How does this walking, shouting, and filling happen in everyday life? Recently we had the privilege of collaborating with communications scientist David Lydon-Staley, who set out to watch humans walk on collective knowledge networks, thereby building personal mental webs.[18] We chose the collective knowledge network of the online encyclopedia *Wikipedia* as it was relatively easy to step upon, being accessible from each person's own living room (or bedroom . . . or bathroom for that matter!). In this study, volunteer participants browsed *Wikipedia* for fifteen minutes a day for twenty-one days. While they browsed, a piece of software—which participants had agreed to install on their computers—sent us information about which sites they visited, in what order, and for how long. The data were then analyzed by treating each *Wikipedia* page as a node in a network and by conceptualizing the transition from one page to another as a step along the edge connecting them. Long steps were those taken between pages with dissimilar content, and short steps were those taken between pages with similar content. The sequence of steps that each person took reflected their characteristic style of curiosity, the kinesthetics of their searching movement, and the building of their knowledge networks.

In the many hours of data collected, we saw clear evidence for two of the archetypal styles of curiosity, the busybody and hunter, and a continuum of stylistic variation between them.[19] Recall that the hunter "wishes [they] had a few hundred helpers and good, well-trained hounds that [they] could drive into the history of the human soul to round up [their] game" in a targeted search for information; in contrast, the busybody "frisk[s] about, and rove[s] about, at random, wherever they please."[20] These kinesthetic signatures of the hunter and busybody should naturally lead to the creation of different individualized knowledge networks. Consistent with this expectation, we found that the more hunter-like participants took walks that traced out networks with high clustering, meaning that they navigated among sets of three or more highly similar pages, taking short steps in a local neighborhood. In contrast, the more busybody-like participants took walks that traced out networks with low clustering and long path length, meaning that they navigated among highly dissimilar pages, taking long steps spanning many distinct neighborhoods. Intriguingly, participants moved somewhat along the continuum between busybody and hunter from week to week, being more busybody-like (or hunter-like) at some times than at others.

Interestingly, a personality trait explained the different network building practices of the busybodies and hunters: their sensitivity to information deprivation. This observation was made possible by data collected in a separate testing session. The participants had completed a scale from the psychologist's laboratory, measuring their deprivation sensitivity by determining the degree to which the following statements characterized them: "Thinking about solutions to difficult conceptual problems can keep me awake at night"; "I can spend hours on a single problem because I just can't rest without knowing the answer"; "I feel frustrated if I can't figure out the solution to a problem so I work even harder to solve it"; "I work relentlessly at problems that I feel must be solved"; and "It frustrates me not having all the information I need."[21] Individuals high in deprivation sensitivity have a drive to eliminate the unknown as they encounter new information and recognize gaps in their knowledge. Consistent with this trait, we found that individuals high in deprivation sensitivity were hunters, while those low in

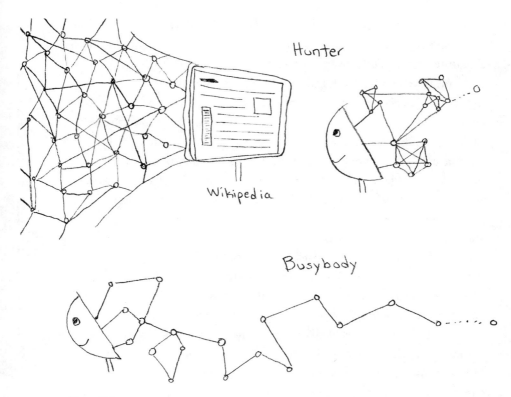

Figure 5.2

Two archetypal styles of curiosity. When browsing *Wikipedia*, the human mind walks on an existing network structure connecting bits of information in the space of collective knowledge. Hunter-like participants take walks that trace out networks with high clustering, taking nearby steps in a local neighborhood. Busybody-like participants take walks that trace out networks with low clustering and long path length, taking long-distance steps spanning disparate neighborhoods.

deprivation sensitivity were busybodies. In other words, individual differences in deprivation sensitivity led to the creation of knowledge networks with distinct architectures.

One cannot help but wonder how Ole Worm would have scavenged in an online knowledge space like *Wikipedia*, and with what authentic responses he would have met the psychologists' deprivation sensitivity scale. What kind of curious was he? Did he set off in a hunter-like search for a specific rare item? Did he roam the world in a busybody-like search with eyes open, ears bent forward, and arms flapping? Did he dance across longitudes and disco over latitudes? In seventeenth-century Denmark, these signature manners of movement feel descriptive and metaphorical, in the sense of abstract adverbs: huntingly, busybodyingly, dancingly. Yet in the modern digital age, they become concrete and even corporeal. They become concrete insofar as the step size of each walk can be measured, the arc of each leap can be mapped, and the choreography of each dance can be written out in symbolic notation as a mind moves (and a mouse clicks) in an online knowledge space. And they become corporeal insofar as the knowledge developed by scientists at the edge of their field is walked, leaped, and danced not just in mind but also in body—in the gesticulations, stance, and pacing that accompany our explanations, whether across the stage, around the conference table, or between the classroom desks. For some, the dance even becomes lucrative. The annual international competition Dance Your PhD is run by the American Association for the Advancement of Science and *Science* magazine, one of the premier outlets to publish cutting-edge and impactful discoveries. The contest challenges scientists working in biology, chemistry, physics, and social sciences to explain their PhD research in interpretive dance. Here the jarring discords of conflicting data and rich harmonies of simple theories blend in the precision of rap, the order of classical symphonies, or the euphoria of blockbuster soundtracks; the conceptual relations and interdependent propositions become movement motifs in ballet, hip-hop, or contemporary dance. The movements of our minds in the process of scientific discovery become the movements of our bodies.

LEARNING KNOWLEDGE NETWORKS

When we as humans seek information, we each seek it differently, thereby constructing fundamentally different networks of knowledge. But suppose that we are not in the state of self-motivated browsing. Instead, we are in the state of most classrooms and institutional meeting spaces, being fed information that we are expected to learn. How does a human learn a knowledge network when presented with a continuous stream of information? The answer can be simply portrayed in a story we told earlier. Recall Robert Macfarlane inviting his mentor, Roger Deakin, to give a talk at the University of Cambridge, and how Roger had provided a lecture whose structure mirrored the structure of the lecture's topic.[22] Roger spoke on waterways, and gave a talk that was watery, in which the topics were connected in a circuitous meander, along an invisible network of tunnels and drains. The story illustrates the point that knowledge networks can be learned by example—the example of how one walks upon them. As a speaker walks from topic to topic, they trace out a network that the listener then sees, ingests, and—if they so choose—owns.

The whole process raises many fascinating questions. Suppose that on a lovely autumnal Friday morning I, as a professor of bioengineering at the University of Pennsylvania, am scheduled to give a lecture on linear algebra, or perhaps curiosity, or even neuroscience. I must choose a set of ideas to transmit to the class, and those ideas are related to one another in a heterogeneous manner, making a network. But I cannot hand the students a network in a single instant in time. Because time is one-dimensional and unidirectional, I must translate that information linearly, passing through one topic before the next, speaking one word before the next. How should I create this information sequence—a set of beads on a string of thought—in a way that maximizes learning? In other words, how do I *map* the network in my mind (which is potentially a high-dimensional object) into a one-dimensional string of concepts traversed in time such that the mind of the listener can optimally *reconstruct* the network's shape?

The mapping and reconstruction processes feel a bit like magic, but somehow humans have learned to perform both easily: every time we communicate, we are mapping, and every time we listen, we are reconstructing. How do we do it? Let's focus on the reconstruction process. The brain of the listener or reader is continually experiencing an incoming stream of data, and from that stream is inferring dependencies between pairs of concepts. By making these inferences over extended periods of time, the listener or reader can eventually infer the whole pattern of pairwise dependencies between concepts. The capacity to infer these patterns is not just relevant to listening to a lecture or reading a book. It is in fact the capacity

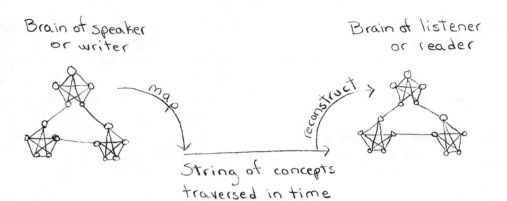

Brain of speaker or writer

Brain of listener or reader

map

reconstruct

String of concepts traversed in time

One word after another ... One line after another.

Figure 5.3

Communicating with one another. When we communicate, there exist at least two minds. The brain of the speaker or writer contains a network, which must be mapped onto the one dimension of time—stringing concepts out one after the other. A person cannot speak a hundred words in the same moment; instead, they must speak one word after the other, and write one sentence after the other. The one-dimensional mapping of the to-be-communicated network must happen in such a way as to allow the mind of the listener or reader to optimally reconstruct what the network looks like. It is only when the speaker (or writer) optimally maps and the listener (or reader) optimally reconstructs that effective communication happens.

that allows humans to learn language, segment visual events, parse tonal groupings, parse spatial scenes, infer social networks, and perceive distinct concepts.[23] So we do it, both commonly and easily. But an acknowledgment of existence is not an explanation of process. To understand *how* we do it, we must design carefully controlled laboratory experiments and use the tools of scientific inference.

BUILDING A LABORATORY FOR NETWORK LEARNING

Can we measure how humans perceive the (potentially high-dimensional) network topology behind a continuous stream of information? Imagine that we stand before a whiteboard upon which we draw a network. We let each node in the network represent an item: a concept, an event, a keypress, or a picture. We let each edge in the network indicate an allowable temporal transition between two items; that is, if nodes i and j have an edge between them, then item j is allowed to directly follow item i in the stream. Next we imagine taking a walk on the network. For example, we could imagine shrinking down to very small (like Alice in Wonderland) so that we could physically step on each circular node and walk across each linear edge (don't look down!). Or we could simply use our finger to trace out a walk on the surface of the whiteboard, along edges, through nodes. Every time we reach a node, with our finger or with our tiny body, we write down the item that the node represents. The process allows us to create a string of items that obeys the structure of the network.

Now imagine that in a separate room sits a human volunteer who has just agreed to participate in our study. We show them the string of items that we have just created. At each item in the string, we ask them to perform a simple task (e.g., press a button on a laptop keyboard) so that we can record their reaction time. The simple task that induces a human response is crucial because the volunteer's reaction time reflects how well they have learned that edge in the network. If they respond quickly, then they have learned that edge in the network well; if they respond slowly, then they have not learned that edge in the network well. This setup allows us to determine how long it takes to learn each part of the network and ask whether some networks are easier to learn than others. In other words, we can ask whether humans can more easily learn networks with spindles like *Sphaerozoum variabile*, or those with meshwork like *Orosphaera huxleyii*, or those with loops like *Octopyle stenozona*.

Before describing recent discoveries, it is useful to recount what had previously been known about the human cognitive processes that are elicited by such a task. What we had known is that humans are sensitive to transition probabilities. For example, consider a person who sees item A transition to item B approximately 75 percent of the time in their environment, whereas they see item A transition to item C approximately 25 percent of the time in their environment. Then when they next see A, they will expect B three times more than they expect C. Because the person is expecting B more, their reaction time to it will also be much faster than their reaction time to C. A corollary to this fact is that a person who lives in an environment where A leads to ten possible outcomes will generally respond slowly due to great uncertainty; in contrast, a person who lives in an environment where A only ever leads to one possible outcome will generally respond quite quickly due to full certainty. Indeed, by placing people in different local environments, we can see that each additional outcome changes the uncertainty by a fixed amount, and every unit increase in uncertainty costs a human an average of thirty-two milliseconds in reaction time. In other words, people process one bit of information every thirty-two milliseconds.[24] To put this number in more concrete terms, it is 0.07 percent of the time it takes blood to move from your heart to your toe and back again (or if you prefer astronomical scales, the time it takes light to travel approximately ninety-six hundred meters, which in turn is about one-tenth of the way from the earth to space).

To understand how people learn the structure of a network, we can keep the local environments fixed (e.g., each node is connected to four other nodes, and every item has exactly four outcomes) while altering the global pattern of edges (e.g., favoring spindles, meshwork, or loops). Together with psychologist Elisabeth Karuza and neuroscientist Ari Kahn, we created distinct strings of items by walking on these distinct networks. We then showed each string to a volunteer while measuring their reaction time to each element in the sequence. What did we find? Humans appear to be sensitive to clusters in the network; they quickly detect the presence of a set of items among which they see many transitions—because in our network, that set of nodes was densely interconnected. We can tell that humans are sensitive to these clusters because they respond quickly to edges inside the clusters, while being surprised (and responding slowly) to transitions between clusters.[25] Interestingly, humans are also sensitive to hierarchy in the network, being more surprised by transitions between or at the boundaries of clusters than transitions deep within a cluster.[26] We can tell that humans are sensitive to this hierarchy because they respond more quickly to items represented by nodes located deep inside the network's clusters, and they respond more slowly to items represented by nodes located at the periphery of the network's clusters.

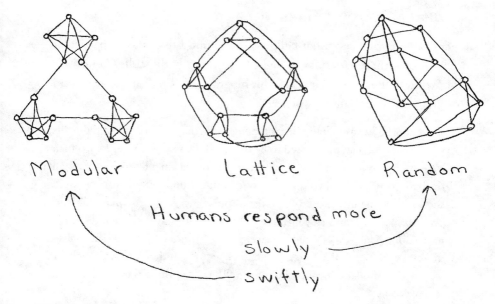

Figure 5.4
Human learning depends on network shape. Humans respond most quickly to sequences of items taken from modular networks (*left*), most slowly to sequences of items taken from random networks (*right*), and middling to sequences of items taken from lattice networks (*middle*).

Experiments of this ilk provide compelling evidence that humans learn some portions of a network better or more swiftly than others. But do humans have a preference for the structure of the entire network? Are some networks overall easier to learn than others? Fascinatingly, humans are indeed sensitive to the overall topology of the full network; they respond most quickly to sequences of items taken from clustered or modular networks, and they respond most slowly to sequences of items taken from random or completely disordered networks. In other words, humans seem to find it easier to learn when information is presented in a modular fashion, which may have implications for optimally organizing a classroom lecture, a piece of writing, or a hands-on task.[27] Collectively, these findings demonstrate that humans can learn a (potentially high-dimensional) network structure from a continuous stream of information. But *how*?

WHAT ARE PEOPLE THINKING?

Understanding precisely what happens in the mind to explain a particular behavior or form a particular thought is notoriously difficult. Cambridge don Arthur C. Benson put it well in his biography of English art critic John Ruskin of the Victorian era when he wrote,

> You know how most of us in idle moments, or perhaps even more in moments when we are officially supposed to be occupied, lapse into a reverie, in which a stream of thought—it may be placid, it may be vehement—sweeps through the brain from the flushed reservoir of the mind. Suppose you check yourself suddenly in one of these reveries. Try to put down in words what you have been thinking of, and as you thought it. You will find it to be ludicrously impossible. Half the thoughts have passed without clothing themselves in any vesture of word, one thing has suggested another, often enough by some trivial similarity of superficial form. The whole thing is evasive, elusive, irrecoverable.[28]

Ack, those unclothed thoughts—they are always the culprits spiriting away our best ideas, refusing to bring back any news from nowhere.[29] But more seriously, what if we were to try to put down in words what people are thinking of as they learn a network from a continuous stream of information, such as from a lecture, a book, a monologue, or some other temporally extended experience?

Perhaps we can gain some insights from Virginia Woolf, who writes in her essay "Why?" of a date on which "in deference to friendship, or in a desperate attempt to acquire information," she had decided to attend a lecture. She recalls, "A large clock displayed its cheerless face, and when the hour struck in strode a harried-looking man, a man from whose face nervousness, vanity, or perhaps the depressing and impossible nature of his task had removed all traces of ordinary humanity." Grim.

> [Then] he cleared his throat and the lecture began. Now the human voice is an instrument of varied power; it can enchant and it can soothe; it can rage and it can despair; but when it lectures it almost always bores. What he said was sensible enough; there was learning in it and argument and reason; but as the voice went on attention wandered. The face of the clock seemed

abnormally pale; the hands . . . moved so slowly. They reminded one of the painful progress of a three-legged fly that has survived the winter. How many flies on average survive the English winter, and what would be the thoughts of such an insect on waking to find itself being lectured on the French Revolution? The inquiry was fatal. A link had been lost—a paragraph dropped. It was useless to ask the lecturer to repeat his words; on he plodded with dogged pertinacity. The origin of the French Revolution was being sought for—also the thoughts of flies. Now there came one of those flat stretches of discourse when minute objects can be seen coming for two or three miles ahead. "Skip!" we entreated him—vainly. He did not skip. There was a joke. Then the voice went on again; then it seemed that the windows wanted washing; then a woman sneezed; then the voice quickened; then there was a peroration; and then—thank Heaven!—the lecture was over.

Apparently, what people are thinking of as they listen to a lecture is "noses and chins, women sneezing and the longevity of flies."[30] How can a person gather an accurate picture of the network underlying a continuous stream of information when their mind ricochets about, losing links, missing sentences, and dropping whole paragraphs? One force draws the mind outward to concentrate on the lecturer's words, while another force draws the mind inward to concentrate on the many questions, preambles, notes, remonstrances, rambles, convictions, footnotes, and postambles bubbling up inside. Could we write down a mathematical model that balances these two forces? Like the equations of motion for a seesaw, or like the trade-off of pushes and pulls in a tug-of-war?

To answer these questions, we had the privilege of collaborating with physicist Christopher Lynn, who drew from the subfield of statistical physics to posit that the human mind works upon the free energy principle.[31] When applied in this context, the principle stipulates that a person's mind seeks to balance two pressures: the pressure to maximize the accuracy of their network model of the world, and the pressure to minimize the computational complexity needed to build that model.[32] The two requirements of maximizing accuracy and minimizing complexity are mutually antagonist, like light and dark. A maximally accurate model of the world would require maximal complexity, whereas a minimal complexity model would produce minimal accuracy. In network learning, the maximal complexity model would provide a perfectly accurate knowledge of the network. Woolf would have savored every link and caught every paragraph, and she would have done so by perfectly controlling her attention. In contrast, a minimal complexity model would lead the human to believe the network is all-to-all connected; in other words, any event could cause any other event, and every item is equally related to any other item. Woolf would have thought much of flies and little of the French Revolution in letting her attention wander; her glimpses of the lecture would feel like random bits of information with equal amounts of dys-relation. "Did a three-legged fly play a key role in the French Revolution?" she might ask herself later that evening.

Each human must strike their own balance in this accuracy-complexity trade-off—a balance between perfect memory and expensive mental effort. Humans appear to enact this balance through a process of fuzzy temporal integration; in their mind, they connect not just two items that appear right next to each other in the sequence but rather a larger set of items that appear near to one another in a wider temporal window. A useful analogy can be made to the perception of a piece of visual art. Instead of focusing on each colored dot in a pointillist painting from Georges-Pierre Seurat or Vincent van Gogh, the mind locally smears the points of color to obtain a Gestalt perception. Similarly, the mind smears points of time to attain a perception of the global structure of a network from a continuous stream of information. As Plato had postulated in his allegory of the cave, humans do not perceive the true world around them.[33] Instead, they see as through a glass darkly, perceiving an inaccurate, smeared picture of their world. Importantly (and contra Plato), though, that biased perceptual process accentuates the boundaries between conceptual categories and other salient architectural motifs in the network, allowing humans to more fully appreciate the large-scale structure of their world.

The model of fuzzy temporal integration is a key candidate mechanism for network learning, allowing us to build the structures upon which we walk and from which we step out curiously into unknown spaces. In the next steps of scientific inquiry, we move beyond simply positing a model by estimating the model parameters from data, validating the model's

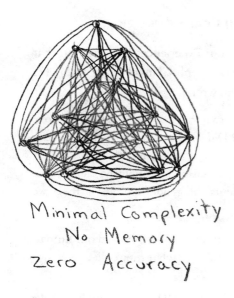

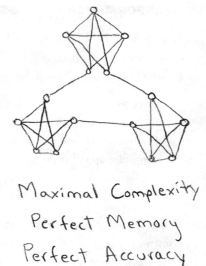

Minimal Complexity
No Memory
Zero Accuracy

Maximal Complexity
Perfect Memory
Perfect Accuracy

Figure 5.5

Humans strike a balance in an accuracy-complexity trade-off, a balance between perfect memory and expensive mental effort. On the left, we see the network built according to the rule of minimizing computational complexity, in which all nodes are connected to all other nodes; building this network required no memory, but it also provides zero accuracy as a representation of the environment. On the right, we see a network built according to the rule of maximizing accuracy; building this network required perfect memory as well as maximal computational complexity. Most humans strike a balance in between these two benchmarks, expending some effort to build a somewhat complex model of our networked world, but also accepting some inaccuracies in that model.

mechanisms, and successfully using the theory to make a prediction about human behavior in a new environment.[34] But even after confirming the model's utility and verity in these ways, the question remains whether the conclusions we draw from the network learning laboratory truly explain the everyday processes whereby we humans communicate knowledge to one another. Laboratory experiments are highly structured so as to control as many of the potential confounding variables as possible. Yet in the process of creating a clean experiment, we risk studying a process that only exists in the laboratory. How do we confirm that our inferences relate to the real world?

INFORMATION PROCESSING IN REAL NETWORKS

Exactly how we build knowledge from our experiences is a question that thinkers have wrestled with for millennia. Joseph Glanvill put it well in his book *The Vanity of Dogmatizing* when he wrote,

> But how is it, and by what art, doth the soul read that such an image or stroke in matter . . . signifies such an object? Did we learn such an Alphabet in our Embryo-state? And how comes it to pass, that we are not aware of any such congenite apprehensions? . . . That by diversity of motions we should spell out figures, distances, magnitudes, colours, things not resembled by them, we attribute to some secret deductions.[35]

When we see symbols, sets of symbols, and strings of symbols in our human communication systems, how does our soul read what they collectively signify, particularly if our perceptions are always biased?

To answer this question, let's consider the systems that humans use to communicate. We communicate using language and music, through social networks and within formal repositories of knowledge. Each communication system is a network. Language networks are formed by concepts that are connected to other concepts according to their semantic relationships or according to their proximity within a sentence. Music networks are formed by notes that are connected to one another by their temporal proximity within the piece. The World Wide Web is composed of sites that are connected to other sites by hyperlinks. Scientific knowledge networks are composed of scientific papers that are connected to other papers by references or citations. Social systems are information networks composed of knowers who are connected to other knowers by friendship, acquaintance, or other forms of collegiality. Collectively, these networks allow us to share information with one another and thus evince the structure of our communication.

Collaborating again with Lynn, we set out to understand what types of architectures these communication systems display, and to ask whether those architectures are sensitive to or resilient against our human biases in perception. In the family of language networks, we studied noun transitions in William Shakespeare, Homer, Plato, Jane Austen, William Blake,

Miguel de Cervantes, and Whitman, while in the family of semantic networks we studied character co-occurrences in Victor Hugo's *Les Misérables* and word similarities in Peter Roget's *Thesaurus*. In the family of music networks, we studied note transitions in Michael Jackson's "Thriller," the Beatles' "Hard Day's Night," Queen's "Bohemian Rhapsody," and Toto's "Africa," alongside Wolfgang Amadeus Mozart's Sonata no. 11, Ludwig van Beethoven's Sonata no. 23, Frédéric Chopin's Nocturne opus 9, no. 2, Johann Sebastian Bach's Prelude and Fugue no. 13 from the *Well-Tempered Clavier*, and Johannes Brahms's Ballade opus 10, no. 1. We also included citation networks in science, site networks in the World Wide Web, and social relationships encoded in Facebook friends and other social media outlets. These exemplars serve to sample each family, and together we used them to obtain insights regarding differences and commonalities across communication systems.

The amount of information held in the network is reflected in how the nodes are connected. The so-called *degree* of a node is the number of edges emanating from it. Networks differ in the number of densely connected hub nodes (high degree) often found in large modules; they also differ in the number of sparsely connected nodes (low degree) often found in small modules. Networks that have a broad range of connectivity degree are said to have high entropy; they pack in much more information than is typically expected in a random network of the same size. The capacity to hold and transmit large amounts of information in short amounts of time is particularly important for systems designed for human communication. In comparing across communication systems, language networks stand out as displaying the greatest entropy, whereas music networks display the least. These differences suggest that real-world networks exist along a spectrum, with some families encoding and transmitting much more information than others.

Notably, each of these network types displays modular architecture, expanded in a hierarchy of large modules composed of smaller and smaller modules.[36] The pervasiveness of modular structure in human communication systems is particularly interesting when we recall that modular networks are precisely those that humans find easy to learn.[37] And why might

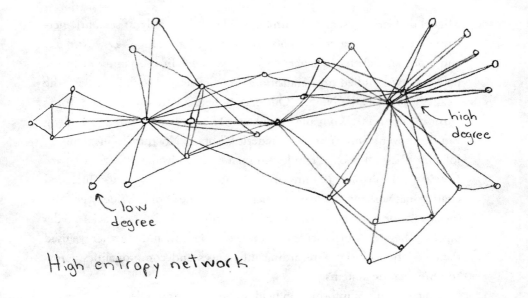

High entropy network

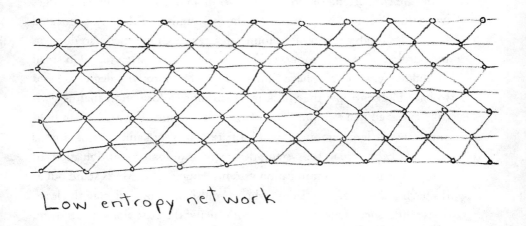

Low entropy network

Figure 5.6

Networks can have high or low entropy. Recall that the degree of a node is the number of edges emanating from it. A high entropy network (*top*) has some nodes of low degree and some nodes of high degree; overall, networks that have a broad range of connectivity degree are said to have high entropy. A low entropy network (*bottom*) has nodes of roughly the same degree; overall, networks that have a narrow range of connectivity degree are said to have low entropy.

modular networks be easy to learn? Their clustered nature minimizes the divergence between the true network structure and the structure that a human perceives through the biased process of fuzzy temporal integration. The idea that some network structures are more or less resilient to human perceptual processes can be quite intuitive when we think again of visual art. Suppose that you stood before a Jackson Pollock versus a Seurat painting; the spatial smearing of your eyes will distill the Gestalt perception of the latter, but only a messy counterfeit of the former. Similarly, networks with certain structures can have their nature shine through our cloudy perception, while others are severely distorted. An optimal communication system among humans should naturally be structured in a way that minimizes the distortion imposed on the perceived information. And amazingly, human communication systems do just that.

But why? Over the past two millennia, have humans preferentially created networks to have architectures that they can most easily understand? Or have human perceptual processes evolved to minimize the distortion of the true networks existing in the world around us? Are networks made by artificial processes (whether intelligent or not) constructed differently from those made by humans? Does knowledge evolve over time to more closely match human network building preferences or vice versa? Laying aside the question of how and why human perception and network structures have become so aligned, it is interesting to ask whether that alignment provides insights into human curiosity. Do our shared perceptions of network spaces allow us to know where to walk for new information and where to strike out into the space of the unknown?

UNDERSTANDING NETWORKS OR THE PRINCIPLES BEHIND NETWORKS?

If networks that house knowledge and those used to communicate knowledge have some shared structure, then perhaps humans learn not just the network itself but also the shared principles of that architecture. And if we learn those shared principles or rules by which the network scaffold spans the space of knowledge, then perhaps we are able to make accurate predictions about where new knowledge might exist or how new knowledge might be structured. A passage from philosopher Immanuel Kant's *Critique of Pure Reason* suggests something similar, albeit devoid of the network phenotype: "Our reason is not a plane indefinitely far extended, the limits of which we know in a general way only; but must rather be compared to a sphere, the radius of which can be determined from the curvature of the arc of its surface—that is to say, from the nature of synthetic *a priori* propositions—and whereby we can likewise specify with certainty its volume and its limits."[38] Perhaps there is enough information in the local structure of a network to determine its placement and span in far-distant areas of knowledge space, just as there is enough information in the local curvature of a sphere to determine its volume and its limits in physical space.

What evidence exists to suggest that humans share an understanding of the principles whereby bits of knowledge are related? Glimmers of a shared perception of the global structure of knowledge can be found in how people understand sets of concepts that are markedly dissimilar from one another. Consider that dissimilar concepts are those that exist far away from each other in a knowledge network, whereas similar concepts are those that exist close by on a knowledge network. Intuitively, it would seem easy for humans to assess the similarity of concepts that are near to one another on the network. But once concepts are far from one another, it might be more difficult to assess their similarity. The intuition, whether right or wrong, is consistent with our perceptions of physical distances. We can quite easily conclude that a ball is one foot away from our shoe, but we find it more difficult to predict how far away a distant mountain is from a nearby cloud. The accuracy with which we predict distances between objects decreases as the distance increases.

Recently, Danielle Navarro and Andrew Perfors—two fantastic scientists who also just happen to be transgender—collaborated with colleagues Simon De Deyne and Gert Storms to devise a set of experiments to understand how humans assess distances over large conceptual spaces.[39] They began by collating 12,428 Dutch words common to the layperson's vocabulary. Within that set, they then chose 300 nouns that had approximately the same word frequency and concreteness, and that—importantly—were not related semantically. They showed adult volunteers the 100 different sets of 3 dissimilar words like "hedge," "belly button," and "peacock," and asked the participants to indicate which 2 of the 3 unrelated words were most related. How might you respond? I ask myself the same question. Should I choose "hedge" and "belly button" because both can be hollo-ways?[40] Or "peacock" and "hedge" because a peacock hedges its bets behind the fabulous tail? Or "belly button" and "peacock" because . . . Well, there really is just no good reason. One might rationally expect that since the words were unrelated, each participant would choose a different pair at random. Shockingly, and almost as if we had a sixth sense, humans in fact consistently pick the same pair from the 3 unrelated words. Why?

The team went on to use several more experiments to show that humans appear to start with a word, and mentally activate the neighbors of that word in the network, and then the next-nearest neighbors, and then the next-next-nearest neighbors, and so on, in a spreading mental activation process. To gain a mental picture of how this happens, imagine that you are standing above a three-dimensional maze, and you pour a bucket of purple paint into one corner of the maze, then a bucket of blue, then green, and then yellow, orange, and red. You watch as the rainbow flows through the channels.[41] But because of the maze's complexity, you do not see a wave of color but instead a jagged progression: in one section of the maze, the rainbow might have spread to the far corner because it was trav-eling along a short straight path, whereas in another section of the maze, the rainbow might still be close to the pouring corner, stuck in a coiled circuitous path. Just as the paint reaches each spot in the maze at a differ-ent time depending on the maze's structure, so every word in a knowledge network is activated at a different time depending on the network's struc-ture. The spreading activation process allows humans to accurately estimate distances and hence consistently choose which pair of words is most related despite the fact that they are far from one another on the network.

The accurate, swift, and shared estimation of distances between con-cepts suggests that humans might do much more than simply learn a network itself. Rather, humans might learn the generic principles of the network's architecture: its shape, its volume, and its limits. We learn what the space is like, what the terrain entails, and what the topography evinces. We learn to expect hills in the Dolomites of X field's scholarship, prairies in the Midwest of Y field's scholarship, and lakes in the receding glacier paths of Z field's scholarship. Perhaps it is this type of learning from which we feel that tug of a pointed guess, the inkling that something beautifully precise exists around the corner of that rock face, the hunch that the next piece of the puzzle should look something like this but not like that. Such tugs, inklings, and hunches may serve to guide our curiosity. We gather a bit of contented happiness when our guess is right and a jolt of enthralled surprise when it is not.

WALKING SAMELY, WALKING DIFFERENTLY

So humans share an understanding of the network structure of knowledge and can accurately estimate distances between concepts as a series of mental steps on the network. Are these shared conceptions the end of the road? Or might they be complemented by idiosyncratic conceptions? Recent advances in natural language processing and computational science give us new tools to answer these questions. In doing so, they have facilitated an even deeper understanding of the network structure of knowledge and the ways in which we can each use it to think curiously.

Early studies began by building finite semantic networks. Investigators would ask participants to perform a simple word association task. As described in an earlier chapter via an anecdote about the professor in T. H. White's *Mistress Masham's Repose*, participants were first shown a word chosen by the experimenter, and then asked to speak the first one or more words that came into their minds. By collating responses from many participants, investigators could estimate the average distances between words. But even more intriguingly, by asking participants to produce such lists, or to produce narratives or stories, scientists could begin to understand the relationships between the produced semantic network structure and mental health or illness.[42] The process of acquiring word association networks from many individuals is painstakingly slow, however, and alternative approaches were needed to scale the investigations. An important complementary approach is to build a semantic network from a written corpus, such as by connecting words that are syntactically dependent on one another or appear within five to ten words from one another in a sentence.[43] The latter procedure builds upon the fact that the co-occurrence of words in a portion of a sentence is suggestive of semantic relatedness under the observation that one is some combination of the company one keeps. Any author's stories, blogs, articles, or books can hence be modeled as word co-occurrence networks.[44] In fact, the approach can also be extended to many authors in large corpora to study how semantic networks have evolved over years, decades, centuries, and millennia.

While powerful, word association and word co-occurrence networks do not provide direct measurements of similarity in meaning but instead only indirect. The most recent advance that offers a fully automatic (easy, fast) and direct assessment of meaning uses machine learning methods to infer latent structures in a large written corpus. A widely used example is Google's Word2Vec algorithm, which distills a word's meaning into a single location in an N-dimensional space of concepts.[45] Just as each person's location on the earth can be specified by two coordinates (latitude and longitude), each word's location in conceptual space can be specified by N coordinates, where N is a number of independent concept features. The Word2Vec algorithm learns the nature of that conceptual space (the nature of meaning) and the location of words within it (their respective meanings). Then distances between the meanings of concepts can be calculated, just as the distance between you and a friend can be calculated by the geodesic distance between your respective locations on the earth. Given the location of words and the distances between words, we can more fully begin to study the ways in which humans walk samely or differently on knowledge networks.

The nature of our walk through knowledge space, along the paths of meaning, can change, adapt, and remodel according to our environment, culture, goals, personality, and mental state. We walk differently through web pages as we browse from day to day.[46] We write differently as we move from draft to draft, for a class project, a book, a journal article, a blog post, or simply an email. We walk differently through new concepts as we

The peacock hedges its bets behind the fabulous tail.

	the	peacock	hedges	its	bets	behind	fabulous	tail
the	0	1	2	3	2	1	1	2
peacock	1	0	1	2	3	4	6	7
hedges	2	1	0	1	2	3	5	6
its	3	2	1	0	1	2	4	5
bets	2	3	2	1	0	1	3	4
behind	1	4	3	2	1	0	2	3
fabulous	1	6	5	4	3	2	0	1
tail	2	7	6	5	4	3	1	0

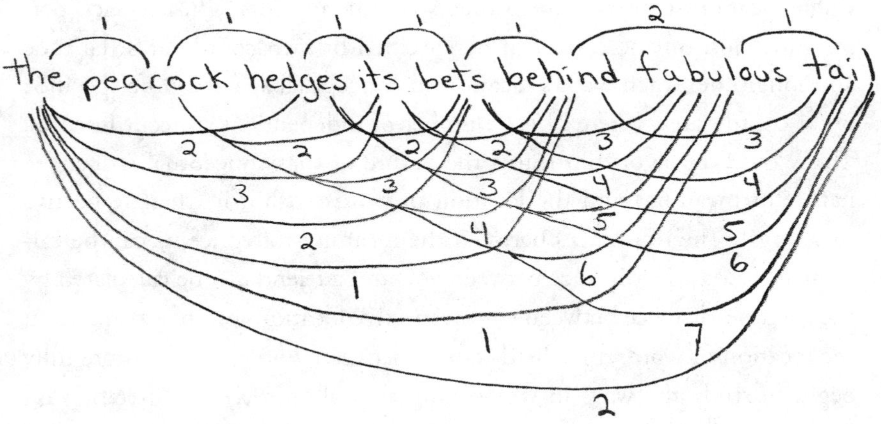

Figure 5.7

Building a word co-occurrence network. Consider the sentence, "The peacock hedges its bets behind the fabulous tail." We can construct a word co-occurrence network from this sentence in two steps using the network scientist's typical canvas. In step 1, place the unique words of the sentence along the rows and columns of the table. In step 2, in the ij-th element (i-th row, j-th column), place the number of words separating word i from word j in the sentence. For example, the word "peacock" is one word away from "hedges" and two words away from "its." (If there are two instances of either word i or word j, choose the instance that produces the shortest distance between word i and word j. The word that occurs twice in this sentence is "the.") For a node-edge depiction of the network encoded in the table, see the lower diagram. Each word is a node, and each edge is weighted by the distance between nodes in the sentence; the weight is depicted by the number atop the edge.

encounter them from chapter to chapter in textbooks and other resources. We even walk differently in different states of mental health, whether through transient states of mild depression or through more permanent states of psychosis.[47] In the former, we may take shorter steps; in the latter, we may take longer steps on the underlying knowledge network, seeing connections where others do not. Walking differently is a privilege that we all have, and one that we can choose to negotiate anew each day. We have the opportunity to adapt our responses to a changing internal and external world that sometimes inspires us to be hunters, and other times inspires us to be dancers.

*   *   *

And in that personal walk, we often use not only a characteristic step length and direction but also a gait, a swinging of the hands, a rhythm of the breath, an emphasis on a word, a euphony of consonant strings or vowel chasms.[48] The movements and sounds we make punctuate the landscape as we traverse the space of knowledge, and from that punctuation arises the cadence of our walk. Our punctuation is a fingerprint, a way of being, a passing through time.[49] The em dash of effort, the colon of climbing, the semicolons of pressing further to the peak, or perhaps the commas of turning about to walk a different path. The periods of pausing for breath before the next summit. The question mark of spinning one's body with arms flung out and face turned to the sky, asking, "Why?" and begging Nature to explain herself. The exclamation mark of dashing the last few meters to peek over the edge to the next valley. Whether we punctuate with our full bodies, or with the movements of our external appendages or internal organs, our articulated walk through ideas belongs to us, and to us alone. Perhaps it is our individual ability to punctuate the air with our waving hands, and orchestrate the stones with our roving feet, that explains why some of us see an opening among the trees while others see a cavern beneath the bushes, and each in their own time steps off the path into the space of the curious unknowns.

6 CURIOSITY TAKES A WALK

> The time of day is less insistent for me than the motion of walking. Trapped in a chair, my mind shuts down. Moving my legs, thoughts start to flow. Swept out of their corners by the flow of my blood, words come into focus, and stories unfold.
> —Anna Tsing, "Writers on Writing" (2015)

In the grip of curiosity, your mind is always going somewhere. Whether scurrying forward, leaping back, or meandering along, your curious mind negotiates existing terrain and refashions it all in one trek. To ask, then, how it is that you think is also, and just as importantly, to ask how it is that you walk.

In the preceding chapters, we analyzed curiosity as a movement in several sorts of ways. We discussed curiosity as traveling along the lines of epistemic relations—or as edgework. We described curiosity as the motivator of knowledge network growth, adding (and reimagining) nodes and edges to existing knowledge systems. We characterized curiosity's kinesthetic signatures, discerning the shapes of its different itineraries as it collects, investigates, or imagines something new. And in the last chapter, we touched upon curiosity as a walk in knowledge space. Here we zero in on *walking* as one kind of curious movement of special significance to philosophy and network science. At least as early as the seventeenth century, network theorists described the traversal of graphic space—representing either conceptual or concrete networks—as a walk. Mathematician Leonard Euler, for example, grappled with the quandary of how to walk across

Königsberg's seven bridges without crossing any more than once. Not only can one walk between places and between ideas, according to specific patterns and within specific constraints, but, as philosophers have insisted for millennia, there is a symbiotic relationship between traversals of thought-space and traversals of world-space. As early as the ancient Greek Peripatetics, the walk has been understood not only as an ideal setting but also as a natural stimulation for philosophical reflection.[1] What conceptual resources, then, does the walk—that quotidian form of perambulation—provide for a network philosophy of curiosity? How might one think of curious thinking as a walk in thought space and of curious travel as a walk in material space? And what does this reconceptualization illuminate about curiosity itself?

Importantly, not all walks are the same, nor are the itineraries of curiosity. As we have argued, curiosity is a pathway, a way of negotiating complex systems, a praxis of knowledge network growth and change.[2] But its pathways diverge, as do its patterns of negotiation. Curiosity strikes off in different directions, donning different shoes, wheels, maps, and snacks. In network science and graph theory, the "walks" taken to traverse nodes and edges have unique structures and constraints. When taking a "walk" simpliciter through a graph, any node and any edge can be repeated. The world, one might say, is your oyster. Trails, circuits, paths, and cycles, however, are specific sorts of walks, each with unique constraints. Trails, for example, can repeat nodes but not edges (e.g., hitting the same peaks but by different treks), while paths can repeat neither nodes nor edges (e.g., hitting only different peaks and different treks). When a trail returns to where it started, it is called a circuit, whereas when a path circles back, it is called a cycle. If curiosity is a walk in a graphological sense, then the question becomes this: What sorts of walks does curiosity take and what constraints, in each case, apply?

In what follows, we take this invitation seriously by attending to how walking produces kinds of thinking and how kinds of thinking are kinds of walking. We investigate four sorts of physical walks, which themselves involve specific forms of curious thought: the philosophical walk, the spiritual walk, the environmental walk, and the political walk. By facilitating different kinds of thinking, these ways of moving through material space map onto ways of moving through conceptual space. One might even call them *curiosity walks*. In attending to these curiosity walks, we characterize the constraints of each sort of walk and each corresponding form of thought. In doing so, we equip ourselves to model and measure these curiosity walks and to imagine their correlative graphs in network theory. Ultimately, we propose that one critical way to cinch together a philosophy and network science of curiosity is precisely by thinking these geographies and graphologies of inquiry together. After all, analyzing curious thought as a walk expands our capacity to tackle the practice and impact of curiosity in new ways.

ASKING, AMBLING

Curiosity and travel have long been thought alongside one another, as if they were companions together along the paths of time. Traveling is a way of connecting the dots, exploring new terrains, while curiosity is a way of moving lightly from one idea to another in a conceptual landscape. At some fundamental level, then, curiosity is a manner of traversing space. Curiosity travels. One might even say that, however much it might seem purely abstract and ideational, curiosity is *spatial*, through and through. The ancients conceptualized curiosity as a movement across spatial boundaries. Whether in imperial expansion, social transgressions, or the literary mixing of marginalia with prose, curiosity was, for them, an untrammeled sort of travel, as deliciously tempting as it was politically destabilizing.[3] The medievals conceived of curiosity as simultaneous travel in two spatial registers: the material world and the conceptual world. For them, curiosity was characterized equally by a wandering mind and a wandering body, the *evagatio mentis* and *evagatio corporis*.[4] As such, the curious person (or *curiosus*) was essentially errant, on the move regardless of the constraints of territory or taxonomy. And of course, there is no era in which curiosity and travel are more synonymous than the modern period, itself dubbed the Age of Discovery and the Age of Curiosity by turns.[5] It is in this moment that curiosity and cartography are most potently intertwined. Curiosity charts a course through strange lands and untraveled seas (with *colonial* curiosity turning mystery into violence and capital). Its very meaning is a certain movement through conceptual and physical space.

Curiosity and walking, as a specific kind of travel, have also been thought alongside and indeed *through* one another. Curiosity is often characterized as a drive to scramble over the horizon, to stride across boundary lines, to traipse around the unfamiliar, and to wander through uncharted landscapes. Walking, in turn, is characterized as a practice of exploration, discovery, and wonder. It is not only the case that one can think of curiosity and walking in interrelated frames, however, but also that they interrelate in concrete ways, such that walking facilitates curiosity, and curiosity shapes where and when and how one walks. Conceptual and physical space exist not simply beside one another in some additive sense but also in relation to one another and therefore mutually affect one another. When one is curious, one is always curious within a specific geographical location. Whether or not one is curious about that geography, that geography is a feature of one's curiosity. More fundamentally, one's curiosity is a form of geographical imagination. Regardless of whether or not one's curiosity is expressed via geographical movement—or where and how one physically moves to investigate something—that curiosity is a way of traversing novel notionscapes.[6] The Latin *solvitur ambulando* (meaning "to solve by walking") refers to a kind of thinking that happens only when walking, whereas the Greek term *methodos* (*meta-odos*, or "concerning the path") refers to a kind of walking that happens only when thinking. Centuries of thought thus invite us to think curiosity and walking together in this robust, relational sense.

Not everyone has equal opportunity or encouragement to walk and to think, however. Indeed, with the rich intertwining of curiosity and walking comes a double prejudice that assumes people of certain genders, races, abilities, or classes are not curious thinkers, nor are they exploratory walkers. As we saw in the introduction, Virginia Woolf was excluded point-blank from the university, where learned men traipsed across the grounds and trailed off into the library. That bias, moreover, saturates centuries both before and after her. For Aristotle, for example, women are not properly thinkers insofar as they—like animals—lack reason. Similarly, for twentieth-century cultural theorist Walter Benjamin, women are not properly walkers. In a shocking passage, he muses:

> For the females of the species *homo sapiens*—at the earliest conceivable period of its existence—the horizontal positioning of the body must have had the greatest advantages. It made pregnancy easier for them. . . . Proceeding from this consideration, one may perhaps venture to ask: Mightn't walking erect, in general, have appeared earlier in men than in women? In that case, the woman would have been the four-footed companion of the man, as the dog or cat is today.[7]

Women, however, are not the only social group to be historically dismissed from the category of true thinkers and walkers. As we move in this chapter to consider philosophical, spiritual, environmental, and political walks, not only do we consciously diversify walking literature (a task, perhaps surprisingly, still deeply necessary), but we also think pointedly about whose curious walks and thoughts get recognized. How might a network philosophy of curiosity break these biases and open up new patterns of recognition in this space?

THE PHILOSOPHICAL WALK

Despite Western philosophy's roots in the ancient Peripatetics, it has not always championed walking as an ideal site for reflection. Today, the harpooning of "armchair philosophizing" is so poignant precisely because its presence is so palpable. And in classroom after classroom, the proverbial "chair" is deployed to support endless thought experiments (from metaphysics to personal identity), inevitably belying the becoming-sedentary of philosophy. That thinking does and ought to happen in a stationary position, however, has ancient precedent. The Greeks worried about the traveling sophists, who were more interested in performance and remuneration than in wisdom. And René Descartes is famous for settling into the chair by his cockle stove to write the *Meditations*. Despite these counterindications, it is not uncommon for philosophers to walk. Jean-Jacques Rousseau would often ramble about in the woods and confessed "je ne puis presque penser quand je reste en place [I can barely think when standing in place]."[8] Immanuel Kant would walk the same route at the same time of day with such regularity that people were said to set their clocks by him. And Friedrich Nietzsche would walk up to eight hours a day, across Germany, Switzerland, and the Mediterranean coast, to court the sort of thoughts that dance in the high mountains.[9] Simone de Beauvoir, too, regularly walked hours on end.[10] Philosophers have been walkers, then. Yet different philosophers and philosophies have assumed different walks, with different spatial constraints. Here we analyze two of the oldest philosophical walks, the Socratic and the Platonic, and their contemporary reverberations.[11] By moving between what Caribbean theorist Édouard Glissant might call archipelagic and continental thinking, we explore two archetypes for the geography of philosophical thought.[12]

Although Socrates fiercely loathed travel abroad, he was committed to traversing—indeed walking—Athens in search of wisdom. Roaming locally, Socrates invited his fellow Athenians to roam with him, in body and in mind. Having heard in disbelief that he was the wisest man in Athens, according to the Delphic oracle, Socrates vowed to assume the work of "wandering [*planē*]" until he discovered the true source of wisdom, outside his own ignorance.[13] This vow led him to a life of wandering—and walking, prey at every turn to the experience of aporia or consternating cognitive impasse.[14] When one routes and rambles, moving hither and thither with no fixed course or destination, one easily stumbles upon the incommensurable and the undecidable, unmasking the rickety scaffolds that support common tenets. For Socrates, this is always more than a private affair. Assuming the mantle of divine calling, Socrates becomes a "gadfly" to his colleagues.[15] While a gadfly incites movement, that movement need not be in any specific direction. Socrates's primary goal is simply to wake the Athenians from their ideological slumber, to shake them out of their epistemic complacency. He goads and guides them into aporia—into the recognition that they do not know what they think they know.[16] The Socratic walk, then, is a collective journey of unknowing. It is inherently deconstructive. "Socrates' wanderings are not acquisitive," Hellenist Silvia

Montiglio remarks; "they lead him to subtract, not to add."[17] Rather than aiming to fill an information gap, Socratic curiosity walks aim to widen that gap, to make it yawn luxuriously.

Unlike Socrates, Plato is a seasoned traveler who nevertheless prefers to sit while thinking. According to the tenets of Platonism, Truth exists in the unchanging realm of spirit. Insofar as the material world constantly moves and changes, it is a force of folly and falsehood. For this reason, the search for Truth must winnow out physical and intellectual wandering in favor of a clear-eyed, sober journey of the soul. Philosophers, Plato writes, "are able to apprehend that which is forever and always the same [i.e., Truth]," whereas nonphilosophers "wander among multiple and manifold objects," and thereby fail to attain Truth.[18] While most of Plato's dialogues open with a wandering Socrates, suspending himself and his compatriots in aporia, they quickly center themselves and everyone takes a seat.[19] In the becoming-philosophical of Platonic dialogues, then, Socrates the gadfly is transmogrified into Socrates the midwife.[20] In midwifery, there is relatively little uncertainty about the course or destination; there is an established plan and an expected outcome. For Plato, then, Socratic wondering and wandering in search of unknowing must ultimately be supplanted by effective dialogue, dialectics, and *diairesis*. *Diairesis* is a method of division and isolation by which things are dissected along their joints.[21] It is by collaboratively conversing in search of finer and finer distinctions that people can attain understanding. As depicted in Plato's allegory of the cave, it is a straight ascent from the darkness of ignorance to the light of Truth. By walking an ever-straighter, thinner line, then, true enlightenment can be achieved and knowledge gained.

Over the centuries, there have been numerous iterations of the Socratic and Platonic walks, but their distinction remains intact. In her documentary *Examined Life*, filmmaker Astra Taylor interviewed philosophers philosophizing while walking.[22] In a manner reminiscent of Socrates, Avital Ronell uses the opportunity of walking around Tomkins Square Park in Lower Manhattan to reflect on philosophy's calling "to make trouble" and "to create skirmishes" wherever it finds itself. Weaving her way through the park, Ronell thinks quickly, responding with agility to the shifting questions posed by her surroundings. Martha Nussbaum, on the other hand, while walking Chicago's lakeshore, launches a well-organized and cohesive argument for why theories of justice should be built not on contracts but instead on human capacities for flourishing. Traversing the relatively straight Lakefront Trail, Nussbaum holds a similarly tight, cohesive thread from beginning to end. Of course, the two modes can also be usefully combined. Judith Butler and Sunaura Taylor, who are interviewed together in San Francisco, conduct a pointed exploration of the social construction of walking itself, but they do so while taking the street and the curb cuts, the strangers and the secondhand shop, as prompts to philosophize. Their philosophical encounter, moreover, is fundamentally relational, and they together invite Taylor's use of a wheelchair to fruitfully derail traditional conceptualizations of the independent, able-bodied walker/knower.

The Socratic and Platonic walks not only have distinct shapes but also house different kinds of philosophical thinking. For Socrates, and arguably for Ronell, anything and everything could be a locale of ignorance. Scurrying about and engaging in random skirmishes is the best way to cohesively disrupt ill-justified epistemic repose. Rather than straightforwardly building knowledge networks, their random walks set about the work of unbuilding, uprooting nodes and orphaning edges. For Plato, and again arguably for Nussbaum, walks toward truth are more somber and staid. Each step must be tested and locked in place, the way one might methodically secure one's footing in unstable terrain. If a path turns out to be a dead-end, one has wandered; one should then return to the main path and parse the terrain more carefully. While this walk may begin with wandering, it becomes ever more keenly focused, with fewer and fewer digressions. Two walks then: one courting fragmentation and opacity, the other consolidation and transparency. Ultimately, whether nodes are added consecutively or haphazardly, and whether edges are heavily reinforced or lightly underlined, philosophy—and indeed *curious thought*—proceeds by way of these paths, inquisitively building knowledge as much as breaking it down to make room for new constructions.

THE SPIRITUAL WALK

Across every religious tradition, spiritual walks have had pride of place. Some walks serve as rites of passage, such as those requiring sojourns in the wilderness. Some serve as daily meditative practices, such as the labyrinth walk, which one finds in religious traditions around the globe. There are the timeless holy pilgrimages, including the Hajj to Mecca, the Kumano Kodo in Japan, and the Camino de Santiago in Spain. And there are walks commemorated across religious history, including the migration of Abraham and the sojourn of the Israelites. While their itineraries have been diverse, however, their functions are quite similar. Spiritual walks, each in their own way, aim to attune the mind, to lend shape to its journey and pattern to its thought. In challenging the mind and body at once, they court connection with the divine. Insofar as spiritual walks seek divine insight and moral transformation, they are curiosity walks. Here we analyze two spiritual walks: the Buddhist walking meditation and the Protestant pilgrimage. While both sorts of walks are millennia in the making, we focus on the contemporary practice of the Buddhist walking meditation and the medieval formation of the Protestant pilgrimage. In doing so, we contrast a walk of arrival, in which every step is on holy ground, with one of destination, in which steps are a means to reach a sacred place. The former is a practice of mindfulness, the latter of purification.

Vietnamese monk Thich Nhat Hanh characterizes the Buddhist walking meditation as a walk of arrival, step by step. Whether one walks to the bus stop, across one's office, in a park, along a river, or up mountains, he writes, "Arrive with every step." Do not walk in order to get somewhere; there is no destination. Invest oneself in each step. Let no part wait eagerly for the next step or peer ahead around the bend. The journey is a single step, each and every step. Steadily bring oneself to awareness, moment by moment. As a practice of mindfulness, such walking cultivates peace and presentness. By it, one slowly sloughs off past regrets and future anxieties. Refusing to run and determined to resist distraction, walking exchanges the "millions of pathways" along which one is pulled for a singular step. Consistent with Buddhist meditation practices in general, the walker is encouraged to focus on what they see and hear, the feeling of their feet touching and lifting from the ground, and the regularity of their breath. "When the mind is focused on breathing and walking," Hanh reflects, "we are already home."[23] Perhaps counterintuitively, this singularity of focus is the precondition of true connection with the living earth beneath the walker and the human, animal, plant, and mineral ancestors within them.

The Protestant pilgrimage, by contrast, is an undeniably goal-oriented, destination-focused walk that, by testing the physical and spiritual endurance of the pilgrim, ultimately secures their redemption.[24] Arriving at the holy place—typically Jerusalem or its stand-in, the pilgrim achieves purification of past sins and thereby a new union with God. The practice builds on scriptural precedents such as the Israelites' four-hundred-year exodus from Egypt to Jerusalem and Jesus's sojourn on earth and return to heaven. In this context, two primary conditions hold: the proper pilgrimage must be arduous, and it must begin and end at home. Walking through unsanctified land, the pilgrim will and ought to experience hardship. This is their testing ground. Neither the mind nor the body should wander. In an attempt to safeguard the pilgrim from worldly pleasures and distractions, the journey begins with invocations of divine aid and pleas for deliverance. The pilgrim is then equipped with special garments, staff, and script, and enjoined to travel the road with somber hymns rather than boisterous tales. Proceeding with eyes half shut and angled toward the ground, the pilgrim walks patiently and purposefully toward enlightenment.

The divergent itineraries of these two walks correspond to divergent paths of thought. The Buddhist walking meditation is quite simple: step, lift, step. Node, edge, node. The walk is over and then it begins again. Mimicking the simplicity of this bodily movement, the mind here ought to focus on an increasingly singular idea: the step itself, holding it in awareness and appreciating its simple components. Uniting step with breath, the walker achieves arrival as a whole being, mind and body attuned. Such thinking is not project thinking but instead present thinking. The Protestant pilgrimage, by contrast, is more elaborate: the walk is long, pocked with trials and plagued by tribulations before reaching the holy place and then returning home. Node, edge, node, edge, node, edge, NODE, edge, node. The mind here ought to be focused on the future, on the hope of achievements unforeseen. Such thinking builds, little by little, so as to realign one's knowledge and revamp one's worldview. It builds both in order to undo certain patterns of thought and to focus one's line of thought on spiritual redemption. The space of arrival (Jerusalem) is a central hub for walkers. All roads, it is said, lead to Jerusalem. Walking to and from this place expands its hubness century after century and heavily weights the edges of certain pathways to it.

Importantly, there are spiritual walks that bridge and move beyond this divide between the walks of project and of presence. As recounted by Robin Wall Kimmerer, central to Anishinaabe culture is the figure of Nanabozho, the first man, and his first walk. Indigenous peoples of the Great Lakes tell the story of Nanabozho, who journeyed to the four corners of the earth to gather its wisdom. He walked to the east to learn knowledge; he walked to the south to learn the secrets of birth and growth; he walked to the north to learn about medicine and compassion; and he walked to the west to learn the interplay of creation and destruction. While each coordinate functioned as a destination, so did each step. Indeed Nanabozho walked in such a way that "each step was a greeting to Mother Earth."[25] Each step was its own journey, with its own hopefulness at the outset and gratefulness at its close. In this way, Nanabozho expressly models a kind of curiosity, a kind of coming to know, that practices presence with the land as much as it undertakes projects upon it. Functioning on two registers or in two dimensions, this curiosity sinks deep as much as it leans forward such that knowledge is built moment to moment, as much as across time and space. Thinking here is communal from the get-go and all along the way, and spiritual wisdom is therefore gained only ever within and in relationship to more-than-human webs of meaning.

THE ENVIRONMENTAL WALK

Although walks have long been taken in natural spaces, nature walks became a substantive practice in the late eighteenth and early nineteenth centuries. From the likes of poets and philosophers William Wordsworth and Samuel

Taylor Coleridge, Rousseau and Wolfgang von Goethe, to Ralph Waldo Emerson, Henry David Thoreau, and Walt Whitman, Romantic walkers assumed an "aesthetic vagabondage" that promised "to heal the self's broken relationship with the world."[26] This tradition extends into (and is reconfigured by) the more recent work of writers such as Woolf and Nan Shepherd, Mary Oliver and Annie Dillard, and Gloria Anzaldúa. Pitched against the alienation of the Industrial Revolution and its consequent mechanization, the nature walk is an effort to reunite self, nature, and world. Leaving urban spaces, and resisting the constraints of road and custom, the ramble is attentive only to the shape of the land and to the habits of its inhabitants. As such, it is a walk of freedom. Rousseau's own saunters are paradigmatic in this regard. As he confesses, "I have never thought so much, existed so much, lived so much, been so much myself . . . as in the journeys which I have made alone and on foot."[27] As testament, Rousseau penned *Reveries of a Solitary Walker*.[28] While de-animating forces swirl among capitalist environments (and their urban ruins), soul and body can be quickened when reconciled with the earth. In this way, one can think new, dynamic thoughts, via renewed, divergent pathways. It is in "the slow respiration of things" that walkers and their thoughts find redemption.[29]

Thoreau, a land surveyor, abolitionist, and essayist, spent at least four hours every day "sauntering." Sauntering is the sort of walking that requires not merely civil freedom but also absolute freedom. Getting the city off one's back, with all of its crusty customs, one must heed only the call of the woods. Refuse domestication. Exchange the front lawn for a bog. And preserve one's inner wildness. In diving into the bush, one must travel *with* rather than against the earth and all of its inhabitants. Thoreau walks "without crossing a road except where the fox and the mink do: first along by the river, and then the brook, and then the meadow and the woodside." Such sauntering, however, is far from mindless. Thoreau's way of walking is always a walking that thinks, that "ruminates." In wandering, he writes, "I trust that we shall be more imaginative, that our thoughts will be clearer, fresher, and more ethereal, as our sky,—our understanding more comprehensive and broader, like our plains,—our intellect generally on a greater scale, like our thunder and lightning, our rivers and mountains and forests." Caught in the confines of society, "the grove in our mind is laid waste." Replanted in nature, however, that grove is allowed to flourish again. Thinking in the open air, while sauntering, allows one to exchange the hay of dead facts for the grass of living ignorance. And then, ever so slowly, this "border-life" permits border-thoughts. Words can be transplanted to the page with "earth still adhering to them."[30] And one can write a book that smells of saplings rather than smudges with library dust.

One of the best-loved US poets, Whitman was likewise a champion of walking. Writing within the Romantic tradition, he naturally preferred to walk on "paths untrodden," away from "the clank of the world," its "profits" and "conformities."[31] Cognizant of both social and intellectual constraints, he often expresses a longing to be "loos'd of limits" and "the holds that would hold me." Scrambling away from "those wash'd and trimm'd faces" of social elites, plagued not only with inner "loathing and despair" but also with "death beneath their breastbones," he seeks to trade duplicity for integrity. Freeing himself from the libraries and literary critics that police appropriate thought-ways, and unshackling himself from the ivory halls and lecture rooms in which empty ideas pass for suggestions of substance, Whitman takes to walking. He pants after the "pathless" universe, yawning and free. He patters haphazardly down its road-ridden face in a last-ditch effort to think otherwise.[32] The worst fate of all would be alienation from the earth and a consequent enervation of the mind and body. Whitman fears suffocation beneath the social mask and walking about as an "unwaked somnambule," incapable of true thinking.[33] Here in the open fields, the thoughts he is given to think are not neatly atomized, as are cities, science, and politics, but rather thoughts of spirituality, of poetry, and ultimately of freedom (and abolition) for all.[34] They are the sort of "large thoughts" found only among great trees.[35]

Among the Romantics' successors, Oliver is a beacon. Hers is an unparalleled commitment to the simplicity of thought and the authenticity of life to be found in nature. A notorious misanthrope, Oliver blithely wanders off into the forest free of society's "smilers and talkers," who are entirely "not suitable" for treks among trees.[36] More of an urban ambler, Woolf nevertheless attributes a similar power to walking in the world outside one's front door. Regardless of where you ramble, even if it be in the wilderness of London's streets, you are likely to have your "shell-like covering" shattered, that sheen by which you manifest the proper person you ought to be, rather than the wild thing that you are.[37] Frustrating that fakery, footpaths lead us into the thickets of our own buried lives, demanding crises of transformation. And yet there is a limit to this freedom and its promise of transformation. As Dillard puts it, you can't "unpeach the peaches." There is always something you carry with you that restricts what it is possible to see and to think. This is deeply frustrating for Dillard. Yearning to nurture in herself the very sensitivity of light, she writes, "The secret to seeing is to sail on solar wind. Hone and spread your spirit till you yourself are a sail, whetted, translucent, broadside to the merest puff."[38] To think with the fugitive freedoms of natural life requires the whispered attunements of the wind—attunements stymied by the social habits and hierarchies that not only saturate cityscapes but also constrain the very reach of nature itself.[39]

While an environmental walk à la Thoreau or Whitman, Oliver, Woolf, or Dillard will precisely not have identical shapes, one after the other, they do have consistent constraints. First, walks of this sort must be taken on a sparse grid, far away from the density of the city, unmarred by the hubs of town or village, or at the very least severed from the tortoise shell of our homes. Second, its itinerary will either creatively reconfigure existing edges, whimsically crisscrossing the network of human roads that traverse the earth, or it will ignore them altogether, following the faintest of creaturely trails and attentively listening to the very shape of the land itself. In both cases, however, this walk will manifest a disdain for the confines of custom and work to unweight the heaviest of edges, reinforced over centuries via social insecurities, racial inequities, economic hierarchies, literary canons, and scientific traditions. One of those heaviest of edges is the very distinction between nature and culture, material space and immaterial thought. Insightfully, Anzaldúa rewrites the environmental walk as always already a philosophical and a spiritual one by which she moves along "el camino de conocimiento," in search of natural harmony as much as insight and enlightenment. From the South Texas deserts to the cliffs of Santa

Cruz, she finds herself "walking between realities" and writing between realities, as only a *nepantlera* could.[40] The work of thinking—of *curious thinking*—here in this space of conch and creature is one of escape as much as engagement, of separation as much as connection.

THE POLITICAL WALK

While philosophical, spiritual, and environmental walks are all political insofar as each in their own way figures and refigures the structure of social organization, the political walk is a thing unto itself. It is a walk whose principal aim is to enact a fundamental critique of existing social structures and to protest the ways in which they compromise human flourishing. These are curiosity walks insofar as they aim to free both mind and society from their current strictures, and to imagine—and indeed embody—other ways of thinking and being. Here we focus on a range of political walks, from the late twentieth-century European *dérive* to the contemporary US protest march. Each is fundamentally in critical dialogue with specific architectural, economic, and social arrangements that jeopardize human freedom. And each not only traverses but also negotiates physical space in such a way as to claim greater intellectual and social space.

Preceding and informing the famous May 1968 uprisings in France, a Paris-based group called the Situationists, led in part by theorists Guy Debord and Raoul Vaneigem, developed the walk of political resistance known as the dérive. Unlike the intellectual act of deriving one proposition from another, which typically leaves little room for fancy, a dérive is a physical movement that creatively contradicts the structural constraints of a given space. It is a walk that pits instinct against institution, playfulness against protocol, and chance human encounters against mechanized interactions. Under advanced capitalism, the Situationists argue, we live in a society of the spectacle, organized around entertainment and simulation, in which we are alienated from reality as much as from one another. Such a systemic "negation of life" is nowhere more present than in our cities.[41] To reopen real lines of participation, communication, and self-realization, it is necessary to stage *situations* that purposefully countermand alienation and reignite vital friction. The dérive is one such reopened line. The French word means to drift, to divert, and to "undo what is riveted."[42] As a practice, the dérive involves a form of "transient passage" that resists inherited urban logics, shifts cardinal direction regularly, and courts the disorientations of chance. It might take the form of nonstop hitchhiking, transgressing demolition projects by night, or "wandering subterranean catacombs forbidden to the public."[43] In staging these situations, the Situationists founded the discipline of psychogeography, which analyzes not only how certain geographical configurations make people think and feel but also how dramatic (if transient) reconfigurations of space change what it is possible to think and feel.

As powerful as the dérive might be, spatial disruption—and its correlative divergent thinking—is not limited to French countercultural intellectuals. Argentine feminist philosopher María Lugones argues that resistant navigation of the physical and conceptual realms is perhaps best modeled by "streetwalkers." Streetwalkers are people who, while they may be homeless and out of work, are at home and at work in the streets. Streets are where these walkers walk not to get out of the house or take a break from work but rather to claim belonging and make meaning. They are border dwellers and boundary breakers. Living "at the border, abyss, edge, shore of countersense," streetwalkers themselves have "ill-defined 'edges.'" They are edge "be-ers" and edge weavers. Streetwalkers walk not in couplets of two or three but rather in multiple clusters, forming and founding "hangouts" as they move along. They hang out where they are not supposed to, when they are not supposed to, and with whom they are not supposed to. And it is in this place of rebellious "mixity" that streetwalkers rethink the world. Busting into spaces, and "bumping among and into one another," streetwalkers generate the "enigmatic vocabularies" that not only illuminate their own lives but therewith defy distinctions of public and private, center and margins, the desirable and the discardable.[44] Streetwalking is a way of thinking, a mode of theorizing, rooted in edge beings and edge makings.

While the dérive and streetwalking protest a whole range of things, they are not protest walks per se. Other sorts of political walks take place in a city—walks not limited to groups of two or three, or clusters here and there, but that extend to tens and hundreds of thousands. Social protests often proceed by way of spatial occupations, in which the function of a space is completely retooled (e.g., highways become footpaths), signifying the necessity of social change on a larger scale. While such occupations may take the form of a sit-in, they may just as often take the form of a walk. Such walks occur most in spaces of preexisting walkability.[45] It is here that political walks not only activate bystander curiosity but more importantly, embody curiosity themselves. They ask the questions: What is going on? What do we need? And what better future can we imagine? One might think of the iconic civil rights movement protest walk from Selma to Montgomery, Alabama. Or the wave upon wave of protests that have flooded the country's capital, agitating for civil rights, women's rights, disability rights, LGBTQ rights, labor rights, environmental justice, and peace. Such walks, moreover, need not include two feet and a salubrious habit of arrested falling. As demonstrated by the 1985 bus accessibility protest in Cleveland, Ohio, alongside so many other disability rights actions, people use their wheelchairs and other assistive mobility devices to walk the city, reconfigure it, and demand social transformation.[46] Each walk, in its own way, proceeds in search of a new world and a new way of thinking.

In the form of the dérive, streetwalking, and the protest march, the political walk resists and rewrites the architecture of a city. In doing so, it redraws the network of streets and social relations; it reconfigures these hubs. For participants and protesters, in changing concrete, they seek concrete change. And that change in social space requires a change in intellectual space. The dérive, streetwalking, and public protest aim to change how people think. How thoughts get linked and configured; how the edges between them are laid and weighted; and thus how questionableness itself is shaped. Of course, there are important differences. The dérive and streetwalking proceed haphazardly, where the path of political protest is largely planned. The dérive critiques the enervating habits of advanced capitalism, while streetwalking and protest typically critique social inequities. In a sense, though, all three carry echoes of the Socratic walk. They are eminently deconstructive insofar as they walk in order to unknow, to undo, and to listen anew. And this is a kind of curiosity given feet on the street.

Travel takes all sorts of forms. One might travel by car, train, or plane. By cruise ship or skiff. By scooter or skateboard. One might run or skip, clomp or clamber. Or one might walk. It is the oldest and simplest form of travel. Focusing on the walk allows for an unparalleled specificity in our accounts of curiosity's kinesthetic movement, especially when those accounts are informed by philosophy and network science. Across the grids of our space-worlds and the graphs of our thought-worlds, the walk cultivates unique rhythms and itineraries by which we come to know our wheres and our wherefores. In patterning those walks, as divergent pathways of traversal, we are able not only to characterize the shapes of concrete and cognitive movement but also explore their interrelations. Just as our bodies and minds are never fully separate, so too are our worldly and wordly walks entangled and intermeshed.

Anthropologist Tim Ingold puts it boldly when he writes, "Walking along is a way of thinking and knowing. . . . [T]he walker is *thinking in movement*."[47] As such, knowledge is not something acquired, or picked up and stashed away in a storehouse. It is not achieved as a destination to which one finally arrives and need never leave. Rather, "knowledge is grown along the myriad paths we take."[48] It belongs to the networks along which one walks and the graphs through which one wanders. Ingold distinguishes between "occupant knowledge" and "habitant knowledge." An occupant enters a space and assumes it for their own externally imposed uses, while an inhabitant finds themselves already in a space and works with other inhabitants to know and to share it. Given this, occupant knowledge assembles epistemic components via categorial distinction and hierarchical organization. Habitant knowledge, on the other hand, pursues knowing horizontally, moving "alongly" across, in, and through the world.[49] "Inhabitants walk," Ingold notes, and hence "their knowledge is not built *up* but grows *along* the paths they tread."[50] For Ingold, as for us, occupant knowledge disavows the implicit pattern of all knowing, which develops along a network of lines and intersections, trajectories and relations. And it is walking that puts the lie to this disavowal. "How, then, does reading differ from walking in the landscape?" Ingold asks. "Not at all. To walk is to journey in the mind as much as on the land: it is a deeply meditative practice. And to read is to journey on the page as much as in the mind. Far from being rigidly partitioned, there is constant traffic between these terrains."[51] We walk—we runners and hikers, artists and scholars, mountaineers and commuters, philosophers and poets, pilgrims and protesters—and as we walk, we come to know.

Cultural theorist Michel de Certeau argues that walking is a way of reading and a way of speaking. There is a grammar and rhetoric to walking. Walking enunciates certain pathways, incorporates certain styles, throws in a flourish here and a digression there. Walking curiously and creatively crafts "the trajectories it 'speaks.'" In doing so, walking is always already a doing and an undoing, a saying and an unsaying. Walking is transformative. In organizing space, de Certeau explains, spatial orders organize possibilities and prohibitions. One can go here, but not there, in this way, but not that. Actualizing possibilities and flaunting prohibitions, the walker's movements are obligatory and transgressive by turns, traipsing down main thoroughfares one moment and "creating shortcuts and detours" the next. Improvising as they go along, the walker redistributes lines of legitimacy and illegitimacy. Networks are reconfigured and graphs redrawn. Such a transformation typically incorporates both synecdochic and asyndetic movements. Catapulting from a part to the whole, a synecdochic walk "leaps" across the graph, inadvertently adding density to the network. Conversely, jumping over dropped conjunctions, an asyndetic walk "skips" around the grid, thinning out the network. Regardless of its traversal tactics (or rhetorical strategies), walking turns surfaces into "sieve(s)," punching holes in the order of things by way of "ellipses, drifts, and leaks of meaning."[52] And it is precisely in this way that walking is a kind of curious thinking, and curious thinking is a kind of walking. Each arranges and rearranges sense.

While all forms of thinking are walks—in both a philosophical sense and a network science sense, curious thinking proceeds by way of curious walks. Such walks, undertaken by seekers of today and yesteryear, shape and reshape the roads of continents as much as those of consciousness. In this chapter, we analyzed four such curiosity walks. The philosophical walk either aims to know à la Plato and therefore plods carefully along, or aims to unknow à la Socrates and wanders about wildly. It facilitates the sort of curiosity that either asks to answer or asks to unmask false answers. In Buddhist, Protestant, or Indigenous traditions, the spiritual walk in turn aims to know either the simplest or the most sacred of truths. It facilitates the sort of curiosity that strives to appreciate the truths everywhere present, or struggles for weeks, months, and even years to attain insights only the most ascetic thinkers can achieve. The environmental walk, as heralded by Thoreau and Whitman, Woolf and Dillard, aims to renaturalize knowledge, leaving behind the confines of common sense and collective prejudice. It facilitates the sort of curiosity that explores paths untrodden, courts insights inconsistent with fanfare, and sticks close to the poetry of the earth. Finally, the political walk, whether in the form of the dérive, streetwalking, or the protest march, aims to know differently and to know freely. It facilitates the kind of curiosity that slips out from under custom and constraint, and, whether pursuing whimsy or justice, it imagines a world in which certain questions are not sacrificed for the comfort of others. Such ways of curious walking are always already ways of curious thinking. And in informing one another, they skip lightly across the lines that have been drawn between ideational and embodied realms. They leap into border-thinking.

*   *   *

When one conceptualizes thinking on a network—walking, skipping, leaping from one idea to the next; building fundaments and orchestrating harmonies; incorporating counterindications or lopping them off; growing and regrowing new nodes and edges, ideas and their interrelations—curiosity becomes something new. It becomes something new and something multiple. What are its shapes, its graphological itineraries? Is it dispersed like the rhizomes of mycelium threads, radial like roots, or intermeshed like the leafy canopies overhead? Or is it a mélange of such patterns stretched far and wide across the globe? What can we learn from network models of thinking in action—and walking in practice—to deepen our understanding of curiosity on the ground? Under what conditions might curiosity behave as a random walk? Or a Lévy flight? A lattice network or a tree graph? But also, what can we learn from philosophy—with its deep appreciation of the divergent thinking made possible by different walks of life—to enrich our understanding of curiosity's creativity (and curtailment) on the ground? What contexts support curiosity—and for whom, and when, and how? Whose mental pathways are recognized and whose sidelined? Whose walking safety is secured and whose ignored? Whose right to thoughtways and footways are extolled, and whose erased? What indeed might Woolf say to Benjamin? In a world where freedom of movement in conceptual and physical space is unequally ensured, how might the capacity for whimsy and wandering, for innovation and imagination, be democratized?

We are excited for future work in both philosophy and network science to formalize these shapes and interstices more precisely, and their freedoms more carefully. There is immense possibility here to illuminate the architectonics of curiosity and to empower the architects. As we toggle between pedestrians and cartographers, and even entertain pedestrians as cartographers of the mind, we can walk into a grounded knowledge of curiosity, situated between disciplinary lines as well as "along" the highways and byways of an ever more mysterious world of words and things. And that story should be everyone's story.

7 YOUR BRAIN ON CURIOSITY

[Some words] show no trace of the strange, of the diabolical power which words possess when they are not tapped out by a typewriter but come fresh from a human brain.
—Virginia Woolf, "Craftsmanship" (1937)

When we are curious, our minds think in different words. Fresh, raw, unused words that well up as the spring rains well down. As we roll them around on our tongues, suck them through the whistling air, and tether them between our teeth—crunchy and uncooked, still green—we realize we've happily exchanged structure for strangeness, and wholeheartedly traded dogmas for daemons. Guiding spirits, they unlock us from the keys, untie us from the knots, and protect us from protection. And so we are at liberty in our own house of curiosity. Here, the Spirit of Intellectual Beauty, for whom poet Percy Bysshe Shelley longed so fervently, is a permanently sheltered lodger, offering "to human thought" its "nourishment" in light, in hue, in consecration.[1] Thus provisioned, our otherworldly power—or more simply our freedom, audacity, and fancy—drives us to dig into, walk across, and leap about conceptual spaces with abandon. But while we'd hoped to revel in secret, to roam in silence, we blush to realize we've inadvertently betrayed ourselves. Each movement we make is prefaced, accompanied, and followed by the sounds of that three-pound, fifteen-centimeters-long piece of tissue inside our skull: shoutings of our inner white bellbird (the world's loudest fowl), whisperings of our inner stout infantfish (the world's smallest vertebrate), and other more cryptic sonorizations of our inner

Nialyod (a rainbow serpent in Australian Aboriginal mythology with the power to transform humans into landscape features; can I request to be a little-known cliff face?). Aside from awkwardly revealing our existence, what do these shoutings, whisperings, and other odd intonations mean? From whence do they arise? How do they create the euphony that is your brain on curiosity?

These questions are only now beginning to be answered. Emerging neurotechnologies can elucidate the mind by taking different sorts of pictures, images, or movies of the brain's structure and function. An MRI allows us to see the spatial layout of different types of tissue and assess the extent to which that layout reflects the contours of a healthy cognitive landscape. This sort of information is akin to the structural plans and specifications of an architect's blueprint, which displays the scaffolding and shoring of the building of our minds. Like X-rays and CT scans in a physician's report, MRIs display the bones, frame, and trusses of the mind's body. But one's body is only part of one's self. Just as an architect's structural plans are complemented by the plumbing, mechanical, and electrical plans, and just as the physician's report may be complemented by an angiogram mapping out arterial flow, so the neuroscientist deciphers the beat, pulse, and meter of the mind using images taken from functional technologies. Functional MRI, electroencephalography, and magnetoencephalography images allow us to probe the metabolic, electrical, and magnetic field signatures of active brain cells. Like a visual artist, the neuroscientist can at times see the story of the mind in the static images, and at other times they must first spool those images together in the form of a time-lapse or hyperlapse to follow the footsteps and track the trails of thought.

The neuroscientist is Dr. Who, using the TARDIS of neurotechnologies to travel to the planet, continent, and country of the mind. And what is this country? And how is the country of one mind different from the country of another mind? The convoluted shape of your wrinkly brain tissue, like the borders of a country, are unique.[2] The thickness of each area of your brain, like the contour lines of elevation on a topographic map, are distinct. The placement of individual neurons and neural populations, like the location of towering cities, is specific to you, as is the location of synapses or wiring, like winding roads connecting the cities. And each of us has a different pattern by which electrical activity carries information throughout our brain, like coursing traffic on highways, city streets, and country roads. Each pattern reflects a string of computations that—unlike those of the most powerful computer—remain a mystery. Identifying what drives traffic patterns of information is one of the key goals of contemporary neuroscience because then we would understand which computations are accessible versus challenging, and which can be learned through careful training or experience. In other words, that traffic pattern is a fingerprint that makes you who you are today and who you could be tomorrow.

The brain is the home country of your curious mind. What happens here when you are curious? To answer this question, we must first ask how the brain is organized. Does the brain share any architectural features with the knowledge networks it seeks to acquire? How do different sectors of the brain relate to different human functions necessary for knowledge acquisition, like motion, vision, decision-making, and—yes—curiosity? When we are curious about a "thing," where does that "thing" live in our brain? When we engage in the edgework of connecting pieces of knowledge, where do the edges lay down their lines? In what bits of tissue are the networks built? Does the neural location of a "thing" along with its relations move or change as we gather new experiences, learn new skills, or entertain new thoughts? (I am thinking of me and my ten siblings piling into our family's fifteen-seat sky-blue Ford van circa 1999: "Move over! There's no more room.") How does the brain configure and reconfigure while we are figuring? The number of questions we can formulate is vast, and the number we know not yet how to ask is even larger. But we must start somewhere. By walking the current trajectory of neuroscientific inquiry, we lay down paths toward an understanding. We begin to see the neural basis of knowledge network building.

RULES OF THE ROAD

How is the brain organized? Many neuroscientists will answer this question by considering the elements of brain tissue at the nanometer scale (one-thousandth of one-fiftieth of the diameter of a human hair). Here ion channels gate the flow of potassium and sodium through the membranes of cells, thereby shaping the brain's electrical signals. The cells themselves span the micrometer scale (just one-fiftieth of the diameter of a human hair). For example, the cell body of a neuron is typically between 4 micrometers (0.004 millimeters) and 100 micrometers (0.1 millimeters) in diameter. The cell body of an astrocyte is typically 10–20 micrometers in diameter. Both neurons and astrocytes have appendages that extend outward from the cell body allowing them to connect with other cells. In neurons, those extensions are axons, which can span up to 1 meter in the human body, and dendrites, which can span from 15 to 1,500 micrometers. In astrocytes, those extending processes radiate out for 20–30 micrometers. As a body extending arms and legs, the corpus-appendage morphology of these two key cell types—neurons and astrocytes—underscores the fact that the act of (and capacity for) connecting is central to brain function. And that connective capacity is undergirded by the impressive fact that the activity of ions in an approximately 50-nanometer region of a cell membrane can affect another part of the body 50 centimeters away through only a single cell. That's a factor of 1 million! (The sounds of the stout infantfish have a very long tail.)

How do neurons connect? To measure neural connectivity, the simplest approach is to extract neurons from nonhuman animal tissue (e.g., from rats or mice) and place those cells in a petri dish with appropriate substrate and nutrient liquid. The individual cells quickly link with one another through synapses typically from one cell's axon to another cell's dendrite, generating a fabric of interconnectivity. That connectivity tends to show a small-world organization where most connections exist between nearby neurons creating densely interconnected clusters, and only a few connections exist between distant neurons allowing electrical activity to span between clusters. Small-worldness is a property that is thought to confer ease of communication across large systems. It can be found in many different sorts of networks for which communication is a central function (e.g., social and telecommunication networks).[3] In fact, it is precisely this architecture, and specifically the existence of the long-distance connections, that allows the parlor game Six Degrees of Kevin Bacon to work.[4] (In this game, players challenge each other to choose an actor at random and then connect them to another actor by naming a film in which both actors appeared; the process is repeated until the prolific US actor Kevin Bacon is reached. The typical number of rounds needed to reach Bacon is six due to the small-world nature of the actor network.)

The small-world feature of network organization is not only present at the cellular scale but also at larger scales.[5] Axons tend to align with one another when traversing long distances and thus the macroscopic organization of brain tissue is composed of tracts, each comprised of many axonal projections. It is as if they find it much easier to walk a path that is trodden than to strike out on a path that is untrodden. In nonhuman animals, tract-tracing techniques can be used to measure these bundles of axons. Whether anterograde (from cell body to axonal terminus) or retrograde (from axonal terminus to cell body), such techniques historically began with the injection of a visualizable tracer molecule into a given area of the brain. The tracer molecules were absorbed by the cells local to the injection and then transported along the neuron's projections. By performing many such injections into different brain areas, a full map of interareal connectivity could be obtained. Across many species, from macaque monkeys and cats to mice and the fly *Drosophila*, tract-based connectomes display small-world properties. The "prolific" areas of each animal brain—the Bacons of neural tissue—are those that have many connections and can easily be reached from other areas simply by taking a few steps on this highly efficient, information trafficking network.

Is the same true for humans, including our friends Ole Worm and Virginia Woolf? The relevant measurements in humans are a bit different from those in animals. Connectivity research in nonhuman animals provides access to cellular and mesoscale measurements, but the same research in humans only provides access to large-scale measurements. Despite the coarser picture, even here the principles of connection architecture appear to be conserved. Using an MRI machine, scientists can acquire a so-called diffusion weighted image, which provides information about the diffusion of naturally occurring water molecules in the brain. Critically, that diffusion is constrained by the placement and orientation of axonal tracts; water molecules diffuse along tracts like raindrops rolling along tree branches. By inferring the location of tracts from the diffusion images, we can construct connectomes for individual humans. The process is a bit like inferring the forest path to the kids' secret fort; like water molecules, the kids stop and turn, chatter and chitter, and zigzag backward and forward, but from their movements and the sounds of their voices we can eventually track the passage under the arms of the forsythia, around the trunk of the hemlock, through the curtain of wispy box elder flowers, and over the old log of a tulip tree. Tracking complete, we are close enough for them to hear us call them in for dinner. When tracking is complete in the brain, we can begin to understand the connectome structure. The largest tract in these human connectomes is the corpus callosum, which is composed of between 200 and 250 million axonal projections. From large tracts to small, the connectome displays a clear small-world structure that is altered by neurological disease and psychiatric disorders, and that changes in normative development and healthy aging.

The clear existence of nontrivial connection patterns in the brain underscores a critical challenge and marked opportunity for modern neuroscience. And that is the precise understanding of how connectivity explains brain function, and by extension, human cognition and behavior.[6] The academic subfield tackling these questions is known as network neuroscience because it combines recent advances in the fields of network science and neuroscience.[7] The commonly shared view is that network organization in the brain is a key component of an explanation of brain function, alongside component parts (regions, neurotransmitters, and genes) and causings (causal processes whereby one part influences another part).[8] Importantly, although pathways (tracts and projections) are not mechanisms, they have the potential to be causes and the conduits of cause, just as forest paths are the conduit for a cause (a parent meandering while calling) driving an effect (kids coming in for dinner).[9] Because of the relations between paths and causes, understanding the pattern of connections in a brain provides an important prerequisite for understanding how information is routed and how computations are performed that support our everyday behavior.

MAKING HARMONY

On April 2, 2013, President Barack Obama announced the Brain Research through Advancing Innovative Neurotechnologies Initiative, devoting significant federal funding over a ten-year period to neuroscience research. The focus complemented efforts in the European Union (Human Brain Project), China (China Brain Project), Japan (Brain/MINDS), and elsewhere. Collectively, that financial stimulus to the field led to the development of transformative technologies and the acquisition of ever more, increasingly accurate, and altogether new data. Efforts have spanned across multiple species and a range of experimental domains, including molecular, cellular, developmental, clinical, and cognitive neuroscience. Pertinent to our discussion here, a key goal in the United States has been to develop a map of the human connectome through the Human Connectome Project. Through these complementary initiatives, it has become increasingly clear that the network architecture of the brain provides simple and intuitive explanations for how the mind works. Why is it that the activity of approximately eighty billion neurons in the human brain does not produce cacophony? Why is it that approximately one-third of our twenty thousand genes primarily expressed in the brain do not create noise? And why is it that the approximately hundred trillion synapses between neurons do not smear information into an auditory milieu that would have delighted Dr. Kakofonus A. Dischord, a scholar of dissonance in Norton Juster's classic, *The Phantom Tollbooth*?[10] To hold at bay (L. *prohibere*) Dischord's purple cloud monster, The Awful Dynne, connections between the component parts of neural systems create precise patterns, allowing the individual voices to blend in a pleasing symphony. The patterns represent not only permission but also prohibition for the flow of information, allowing concordance and harmony.

To support harmony, the connection patterns in neural systems are more than simply small-world ones. They are modular. Modular connection patterns provide dense conduits among a particular set of brain areas that collectively support a given function.[11] Intuitive examples of modules are those that support our sensory and motor abilities. Regions of the brain that support body movements are densely intraconnected in a motor module; similarly, those that support vision and audition are connected in visual and auditory modules. Like the string section in an orchestra, the many parts (musicians or brain regions) work together through connection conduits (of shared melody or physical wiring) to create the music of the mind. Beyond motion, vision, and audition, brain modularity also explains complex cognitive capacities that include one's understanding of one's self and one's ability to engage in complex reasoning. Regions of the brain that support working memory, inhibition of inappropriate behavior, and the making of difficult decisions are densely intraconnected in an executive module; regions of the brain that support the voluntary deployment of attention and the reorientation to unexpected events are densely intraconnected in an attention module; and regions of the brain that support self-reference and self-referential processing are densely intraconnected in a default mode module. The regions of the latter module also tend to be highly active at rest, suggesting that the modus operandi of humans is, at least in part, one of contextualizing experiences in relation to the self.[12]

But how do modules communicate with one another? How do I hold a leaf in my hand and note the curious branching of its veins (visual module); then reflect inward to the blood coursing through my own veins, appreciating the similarities between the leaf and I (self-reference module); and then use that appreciation to turn back to the leaf to study its vasculature in even greater detail (attention module), for which I move my body to get a closer look (motor module)? It is clear that even the simplest of curious experiences requires a coordinated flow of information between modules and a masterful orchestration of that flow (executive module). This coordination supports simple curiosity, yes. Yet more than that, it supports basic survival as well as the ability to pursue my goals and gather various resources for all the other dimensions of well-being. The interdigitation of cognitive processes—from curiosity to sincerity, from temperance to authenticity—is supported by connections that link one module to another, ensuring that each process is not fully encapsulated but instead naturally tethered to others.[13] The delicate balance of dense modules and lenient fastenings among them allows a system to optimally adapt or evolve by stretching, strengthening, or shifting one module without perturbing the others.[14] These adaptive affordances of modularity are thought to explain the marked fine-tuning of intermodular connections that occurs as the brain develops, and the fact that modularity can be used to predict how plastic the brain will be in response to cognitive training, aerobic exercise, and other activities.

Modularity helps to explain the process or practice of curiosity. But several questions remain. As reviewed in chapter 1, evidence suggests that curiosity is supported by two circuits: the reward and motivation circuit, and the cognitive control circuit. The former is active during curious search and sampling, partially reflecting both the motivation to obtain novel information and the feeling of reward when that novel information is obtained.[15] The latter controls the timing and nature of both search and sampling, in part by monitoring whether cognitive resources should be maintained toward the current goal or redirected toward other potential ones. Neither circuit has a perfect one-to-one mapping onto the modules we just described. Instead, the two circuits span among those modules, indicating the existence of a coordinated—sometimes sequential, sometimes parallel—function, allowing us to engage in the complex process that is curiosity. Curiosity is like a hand interdigitating the modules of the brain into a palm of mental play. Or more than that, curiosity is the contours of the fabric we weave when we pull on the string of one lightly tethered module and slacken the string of another lightly tethered module. Curiosity is the skilled hand of a worker at the local party shop, patiently tugging ribbons to align the balloons into a shapely bouquet as if it were some silent, buoyant, colorful thicket of thought.

WHERE IS A SINGLE MELODY?

In the *neural* processes of curiosity, we see reflected the *conceptual* processes of curiosity. Neural circuits supporting curious behavior reflect a temporal path through a series of lightly tethered modules of the brain. As the mind moves from one local function to another by traversing a network of interconnected brain regions, we are reminded of the human mind moving from one concept to another by walking upon a network of interconnected pieces of knowledge. In other words, it is as if the neural mechanism of curiosity in a single human's brain network reflects the mental process of curiosity across the collective humans' knowledge network. Our electrical signals walk on networks, as much as our minds and our bodies walk on networks. The mirroring between brain and mind motivates a deeper understanding of each step, and not only of the space traversed by the step but also of the location on which the foot is planted while the gait is adjusted for the next movement. In the physical realm, the foot is placed on a bit of earth; in the conceptual realm, the mind is placed on a bit of information; and in the neural realm, the activity pauses in a bit of tissue. That pause underscores the importance of understanding not just the collective role of all regions in a module or circuit (like the wind section of an orchestra) but also the singular role of each region in the process (like the melody carried by the English horn). Does the bit of tissue in the brain reflect the information in the mind, which in turn reflects the bit of earth in the world? And if so, how?

A bit of brain tissue can indeed hold within it a picture of what the eye sees, the hand holds, or the feet feel. That picture is commonly referred to as a representation and is typically comprised of a pattern of activity across neurons or groups of neurons.[16] Imagine a pixelated image in gray scale; each pixel is a neuron, and each neuron has a level of activity (or rate of firing) that determines the shade of gray that pixel contains in the image. The image can be deciphered, the meaning can be decoded, and the mind can be read by first learning the relation between the neural and real-world images. We are still learning which images are best deciphered where. Some bits of brain tissue have a tendency to provide pictures of particular items in our world. For example, there exists a swath of tissue that tracks along the bottom of the brain from the back to the center, and contains subregions whose pixelated image of activity encodes first shapes (furthest back; lateral occipital cortex), then word forms (fusiform gyrus), then faces (fusiform face area), and then places (furthest forward; parahippocampal place area). Is there also a focal point for harmony? A bit of earth for humicubation?

Mind reading—or more accurately, brain reading—is a form of decoding. It is the capacity to predict what you are seeing from the pattern of neural activity in the brain. While it sounds a bit like magic, it is closely related to the reverse operation: predicting the pattern of neural activity in the brain from what a person is seeing. In other words, by studying the relation between the neural and real-world images, we can infer how the brain encodes the world. We can begin to conceive of the cipher and crack the code. What are the rules whereby the brain transforms experiences into electrical signals? And which parts of those experiences (or the images they contain) does a given human truly "see" or actually process? We know from simple everyday events, like sitting by a three-year-old and together looking up at the cloud formations in a summer sky, that each of us perceives the world differently: "I see an elephant holding a candy cane!" "Where? Oh. I thought it was a mug of coffee on a bed of romaine. Your elephant is much cuter than my coffee." By studying how real-world images are encoded in neural images, we can begin to understand which bits of the cloud became flora for one observer, and which bits of the cloud became fauna for another.

A real-world image need not be represented solely in one brain region. Instead, experimental evidence indicates that different parts of the same image can be represented in different parts of the brain. For example, an object's identity might be encoded in a bit of temporal cortex, its shape might be encoded in a bit of lateral occipital cortex, and its color might be encoded in a bit of visual cortex. That fact allows neuroscientists to start to understand how the whole relates to the parts. Are the parts simply added (or multiplied) to become the whole? Are some parts given more say than others in how the whole is represented? (Woolf likely wouldn't be surprised.) And where do the parts get combined in the brain? Scientists are beginning to build evidence for a "convergence zone" where some kinds of whole-making happen and the true concept emerges. Moreover, we are starting to see how information encoded in one brain region is shared with another brain region as well as how that information is transformed or changed in the process of transmission.[17] From these efforts, we hear the melody of single brain regions and learn how they join together to form the symphony.

WHERE DO WE BUILD A FULLY MUSICAL LANGUAGE?

If a harsh sentence from the judges awaits someone, once
He has been condemned to afflictions and penalties,
Let workhouses not fatigue him with raw material to be wrought
Nor let mines of metal pain his stiffened hands:
Let him make dictionaries. Need I say more? This
One labour has aspects of every punishment.
—Joseph Justus Scaliger, *Poemata* (1615)

A philologist, historian, and religious leader in the Renaissance, Joseph Scaliger could think of no more laborious sentence of penal servitude than to make dictionaries.[18] The irony of his scathing opinion, as evident to a modern ear, lies in the fact that the human mind—every human mind—makes a dictionary. That dictionary is housed in the spherical confines of the brain—the shape being a biological mirror of the curated collections housed in the circular building of the Bibliotheca Alexandrina, the circular dome of the Bodleian, or the circular reading rooms of the Bibliothèque Nationale de France. And what is a dictionary? A dictionary is a collection of words, a placement of words, and a "definitionary" of words. How does the mind collect, place, and define words? We collect words and the concepts behind them through the natural processes of language acquisition starting as infants and, for many, continuing into old age.[19] And as we have just seen, we place words, concepts, and features of concepts in particular bits of brain tissue. As Woolf notes, "Words do not live in dictionaries; they live in the mind."[20] But where are the words in the mind? Where are the definitions? The cross-references? The clarifications of similarities and differences between words? Answering these questions requires us to dig more deeply into how, where, and why concepts and their relations are represented in the brain.

Imagine you are lying down in a dark room, covered in a warm blanket, hearing the comforting bang, clang, rattle, and thump of the MRI machine engulfing your head and upper torso. (Don't worry. If you have claustrophobia, you were already excluded from the study.) Above the sonorous serenade, you hear an episode from the Peabody award-winning

Moth Radio Hour, a weekly series produced by PRX and Jay Allison of Atlantic Public Media, airing on over 450 radio stations, and featuring true stories told live onstage without scripts, notes, props, or accompaniment.[21] Every second or two, a forty-thousand-pixel snapshot of your brain's activity is taken. Scientists carefully compare the time at which each word of the transcribed story was spoken and the time at which each pixel of your brain image showed activity. From that comparison, the scientists build a computational model that predicts your brain activity from each word you heard. To determine the model's validity, they also predict what patterns of brain activity you will show when you hear a new story (imagine Margaret Cavendish's *The Blazing World*, François Rabelais's *Gargantua and Pantagruel*, or Jonathan Swift's *Battle of the Books: A Full and True Account of the Battle Fought Last Friday between the Ancient and the Modern Books in Saint James Library*); then they have you listen to that new story and test the accuracy of their predictions.

This experiment became a reality for several adults living in the San Francisco Bay Area.[22] From their data, University of California at Berkeley scientists could visualize what semantic information was stored in every piece of the cortex. What they saw was an incredibly complex semantic map that tiled the entire brain, and not just areas that had previously been implicated in semantic processing. Every millimeter of brain tissue was associated with a word, and similar words tended to elicit activity in nearby pieces of tissue. Sometimes the same bit of the brain could even activate to more than one word, albeit differently. Returning to Scaliger, we realize that the mind's dictionary lays out its curated collections across the entire real estate of the brain; no millimeter is left uncovered. This library has not converted half of its footage to other uses (the café, the computer lab, the atrium). No. In the country of the mind, libraries devote their entire structure to the stacks. And even on the same bit of real estate (the same tissue section), the mind stores many different words, just as a library stores many volumes on a single bookshelf's footprint or a dictionary stores many entries on a single page's column.

But beyond a collection of words and a placement of words, the mind is also a definitionary of words. Where can the definitions be found, and how are they read? Using analogous experiments, scientists can measure the brain activity evoked by one word and calculate its similarity to the brain activity evoked by another word.[23] By measuring this similarity between many pairs of words, we see that the brain produces similar activity patterns to certain sets of words, and each set thus forms a brain-defined category reflecting the brain's encoding of meaning. The *Oxford English Dictionary* may categorize words of the language into scientific, foreign, literary, common, technical, slang, and dialectical; British nature writer Robert Macfarlane may categorize words of the UK landscape into flatlands, uplands, waterlands, coastlands, underlands, northlands, edgelands, earthlands, and woodlands; a given human brain might categorize words into animal, vehicle, natural objects, and so on, as well as into subcategories such as mammals, boats, and grasslands.[24] (Scottish poet Norman MacCaig would be particularly proud of that last subcategory after penning the following in 1983: "Scholars, I plead with you, Where are your dictionaries of the wind, the grasses?"[25] This brain, at least, is a scholar that does not disappoint.) The categories that the brain "sees" in this experiment are far richer than previously thought.[26] Because the categories provide broad strokes information regarding words' meanings, they may differ across cultures and eras, societies and times, bodies and experiences.

The precise meanings of each word can be read in the pattern of relations among categories, subcategories, and exemplars (or single words). That pattern itself creates a network replete with complex branching arrangements and hub nodes with high degree, reflecting words that share similar patterns of neural activation with many other words. In one visualization of that network, the categories and their relations inferred from brain activity form a treelike structure reminiscent of the phylogenetic trees of German anatomist Maximilian Fürbringer: then for forms of flight, now for flights of fancy.[27] As we walk curiously through the mind's conceptual space, we plant our feet on bits of brain tissue that code each word of each thought; we meander across the hills and valleys of the brain's sulci and gyri. Like writer William Joyce's bookish character Morris Lessmore, we cavort across lines of anatomy, swing from trailing J's of curvature, and slide down the creased binding of the dictionary of our mind with a full language in hand.[28]

WHERE ARE NETWORKS CODED?

Why are there trees I never walk under but large and melodious thoughts descend upon me?
(I think they hang there winter and summer on those trees and always drop fruit as I pass).
—Walt Whitman, "Song of the Open Road" (1858)

Who penned our dictionary? Does it have a creator equivalent to Robert Cawdrey (1604), Samuel Johnson (1755), George Barrow (1873), James Redding Ware (1909), or even our friend Ole Worm, who published the first collection of Old Norse vocabulary in 1636? Or did the words, like thoughts large and melodious, fall from trees as we passed them on our walks?[29] As far as we know, the creator of our word list is Nature, who writes with the pen of experience and the ink of time. Experience—shot through with the social and cultural—provides us with a conceptual space and builds the architecture of that space. Where do we house the architect's floor plans of our conceptual spaces? And how do we use those plans to navigate the conceptual world? To think the thoughts we do? To eschew the thoughts we do not?

To answer these questions, it is useful to first consider how we build maps of physical rather than conceptual spaces. The process of mapping our physical world seems simple and is certainly more well understood than the process of mapping our conceptual world. Moreover as we shall see, mapping physical spaces has marked relevance to mapping conceptual spaces. The mapping largely occurs deep inside the brain in an area called the hippocampus, which spans approximately three to forty centimeters, thus making up 0.25 percent of the brain's total volume.[30] Tiny. Here a particular type of neuron exists called a place cell.[31] A given place cell fires whenever you stand in a particular spot in your room, home, or workplace, in a field, or on a mountain. As you stand in each spot in a given space, a different place cell fires. But place cells are not the sole mapping mechanism in the brain. A neighboring region, the entorhinal cortex, is only a fraction of the size of the hippocampus. Tinier. Here a different type of neuron exists called a grid cell.[32] Grid cells create a map of a given space by tessellating that space with equilateral triangles, thereby building a periodic hexagonal lattice. Together, place cells in the hippocampus and grid cells in the entorhinal cortex allow the typical human to navigate true physical spaces, allow the atypical human (a seasoned taxi driver) to expertly navigate the London streets, and allow the (hopefully not untypical) fearless spirit to travel in the mind.

Building on these discoveries, scientists have sought to press beyond an understanding of how we navigate space, either terrestrial or celestial, and toward an understanding of how we navigate conceptual spaces. They have wondered, guessed, and even attempted to predict how the brain engages with, computes about, and processes the spaces in which concepts live. Common among those thoughtful engagements is the notion that the brain organizes concepts into a mental map in the same way that it organizes locations into a physical map. As the brain codes the nature of physical spaces with a grid, so might it code the nature of conceptual spaces with a grid. If true, perhaps the same principles that allow us to navigate physical spaces also allow us to navigate conceptual spaces: landmarks, distances, and available pathways. Recent work in humans provides precise evidence for just such a grid-like code that organizes our abstract knowledge in a two-dimensional conceptual space.[33] What is particularly beautiful about the code is that it exists not just in the entorhinal cortex where it has been observed in other mammals but also in other areas of the brain implicated in memory, imagination, and theory of mind, and in areas that are active as we navigate new situations with tools we have acquired in previously experienced situations. In other words, our memories may be mapped, our imaginations may live in a dimensional world, our conceptions of others may use carefully measured notions of distances, and we keep these maps in the back pocket of our cargo pants as we explore the vast lands of new experiences.[34]

Collectively, these data suggest that the space of the mind is not so ethereal as we might have previously imagined but rather is tangible, walkable, climbable, and even leapable. How do we learn the pathways to draw on the map? How do we lay down the sunken lanes of Southern England, the *chemin creux* of Northern France, the *corredoiras* of northwestern Spain, the holloways of West Wales and Syria, the Kaidō of Japan, the sunken Natchez Trace of North America, and the well-worn paths of our minds reflecting the truths about ourselves and our world? In essence, these questions amount to the scientific question of how we learn knowledge networks, the pathways that define the relations among things ("outside these relations there is no reality knowable," as we recall from Henri Poincaré).[35] An especially useful experimental approach to answering this question is to show humans a string of objects in which certain objects tend to follow certain other objects, as we discussed in chapter 5. Unbeknownst to the human observer, the string of objects comes from a pattern of relations (or a network) between the objects, with related objects often being seen in series and unrelated objects never being seen in series. Using such a network learning experiment, scientists have uncovered evidence that the hippocampal-entorhinal system sees the associative relations among objects and codes them as distances.[36] Those distances are mapped in two ways. In the first mapping, a bit of entorhinal cortex codes whether there exists a path (a road or relation) at all between the two objects in this conceptual space. In the second mapping, a different bit of entorhinal cortex codes, for disconnected objects, how many paths (multiple roads or relations) must be traversed to get from the first object to the second object in this conceptual space. Together, these two codes make a maplike knowledge structure that lies in the dark of our subconscious, lighting the way as we walk our world.

FLEXIBILITY AND LEARNING

The Scottish poet, mountaineer, and naturalist Nan Shepherd writes of hours of walking in the Cairngorms, noting that "the eye sees what it didn't see before, or sees in a new way what it had already seen. So the ear, the other senses . . . These moments come unpredictably, yet governed, it would seem, by a law whose working is dimly understood."[37] As we walk in the landscapes of the natural and conceptual worlds, we are often struck by new perceptions, add distinct haunts, forge a different path, or change an old one by stamping and splashing disobediently down the middle of the carefree rivulets after a rainstorm. Landscapes change, as does collective knowledge. Paths change, as do the relations among bits of knowledge. Our networks change. As we age from lithe youth to calmer elder, the very manner in which we walk changes. The maps change, and the way we use the maps changes. Different maps accompany structures of oppression, while acts of resistance change the maps and reclaim their use. What rules govern all of this audacious changeability?

The brain needs flexibility. It needs flexibility to learn as new data arrives, to morph as the body morphs, and to change as the world changes. Yet much of the brain—its structure and even its function—is hardwired or relatively fixed. We do not grow new brain areas during our lifetime; the large axonal tracts described earlier in this chapter maintain their arrangement throughout adulthood. Hence perhaps our greatest hope can only be for the brain to move as swiftly as a cheetah running through molasses, or as adroitly as a tree attempting to outpace the cheetah in a never-ending cycle of uprooting and then replanting, uprooting and then replanting. How does the brain solve the quandary of flexibility within (the admittedly massive) constraints of inflexible hardware? Recent evidence suggests that it does so in part by traffic control: changeable stop signs, editable traffic lights, inflatable (and deflatable) speed bumps, and so on. The brain carefully orchestrates the manner in which information is shuttled along connections (footpaths, roads, and highways) and particularly along those connections that bridge one module to another.[38] Brains with more flexibility of intermodular communication learn more effectively than those with less intermodular flexibility. Moreover, the degree of modular flexibility in a person's brain can predict their capacity to learn in the future. Such flexibility is modulated by sleep, mood, and mental health, and fundamentally allows for the brain to adapt without reforming.

As a mechanism for freedom, flexible communication between modules and circuits could allow us to build, use, and revise our cognitive maps of knowledge. Connecting to the attention circuit could determine which pieces of the world to attend to or perceive (the elephant holding a candy cane or the mug of coffee on a bed of romaine), and which pieces to subsequently draw on the map in our back pocket (the hippocampus-entorhinal system). Connecting to the reward circuit could determine whether using the map was a positive or negative experience, influencing which maps to continue building and which to lay aside. Connecting to the motivation and valuation circuits could determine when to use the map, and how much to value the map of one cognitive space (math?) versus another cognitive space (theater?). Connections to the motor circuit could determine how the body moves in new ways (walking, running, leaping, dancing, athletics, or playing an instrument) to find fresh cognitive or emotional places on the map or novel roadways to lay down. Connections to the visual and auditory circuits could determine how we see and hear each part of our world, and encode it in our conceptual space.

Flexible connections can allow information to be shuttled differently between modules or circuits, and in turn that flexible shuttling can change how we represent the world and the entries we place in our dictionary. Words commonly change their meaning over time in response to changes in society, culture, science, and many other factors; moreover, the concepts behind the words can also change. For example, consider the word *imply*. In the fourteenth century, it meant "to enfold, enwrap, entangle"; the meaning "to involve something unstated as a logical consequence" was first recorded circa 1400; and that of "to hint at" is from the 1580s.[39] Or consider the word *impertinent*. In the fourteenth century, it meant "unconnected, unrelated, not to the point," but the sense of "rudely bold, uncivil, offensively presumptuous" is from 1680s, from the earlier sense of "not appropriate to the situation" (1580s). The changeability of words implies impertinence, a certain presumptive freedom from the rigorous constraints of fixed meanings. That changeability is a celebrated feature of words in any language. And the language of the brain is no exception. The meanings of words in the brain change as the neural activity representing those words changes. Representations (patterns of activity across neural units) evoked by items in our world can strengthen, crystallize, and adapt as we go through life. Although not yet well understood, such changes in neural representations could be made possible by the flexible, different, and adaptive routing of information throughout the network of the brain.[40]

THE FLEXIBILITY OF PRACTICE

The flexibility of our brains allows us not only to change and grow our understanding of the world around us but also to engage with the world ever differently. It allows us not only to be curious but to practice curiosity, to change curiously, and to adapt our curiousness. What does it mean to practice curiosity? Is curiosity something that can actually change and that can truly grow? Many scientific studies operate under the assumption that curiosity is an innate or default state: a capacity that is best characterized as a trait of a person or the common mode in which the person operates.[41] This notion is similar to the one that a person has a natural level of bombastry, bookishness, or brilliance. Yet the evidence suggests that many dimensions of our personality and personhood can vary over quite short timescales. Let's consider the central human capacity for mindful attention. Far from fixed in a single person, mindfulness can be demonstrably altered by training, thereby fundamentally modulating a person's patterns of thought, leading to a change in their decision-making, working memory, spatial memory, verbal fluency, and cognitive flexibility. And all of this change making is possible because mindfulness alters the activity of specific areas of the brain. Similarly, new data indicate that curiosity is far from fixed in a person; instead, it can wax and wane naturally from moment to moment. Further, curiosity can be modulated by external factors including those present in learning environments. The fact that curiosity can vary and be varied opens up the possibility of practicing curiosity with the aim of self-betterment. Perhaps it also opens up the possibility of practicing curiosity alongside other humans for the betterment of our society.

Practice is a form of training. In mindfulness training, one practices a certain set of mental states and a set of transitions (or lack of transitions) between them. In curiosity training, one might practice mental states of curiosity and mental state transitions following a line of inquiry. Moreover, one might practice choosing the objects of curiosity, following patterns of curious search, and making time and space in one's life to act upon one's information-seeking proclivities. Metaphorically, one walks along one's network of knowledge, and seeks to build new webs, add new edges, append new nodes, or leap into the black (or blank) space beyond one's knowledge in the hopes of landing on some deliciously unexpected idea. The manner in which we walk, build, and leap may be informed by our personalities, educational experiences, learning capacities, and predilections for hunting, busybodying, or dancing. What types of nodes do we seek? Are they single concepts or larger ideas? What type and distance of links are we willing to make? How distinct may two ideas be for us to still acknowledge their relationship? What is the manner in which we incorporate the new node and edge into our existing knowledge network (if at all)? What sort of architecture do we wish to build? Is it dense or sparse? Ordered or disordered? Low dimensional or high dimensional? A wastebasket or jewel casket? The answers to these questions require some explicit notions of distance, geometry, and space: the distance between ideas, the geometry of the network, and the space in which the network exists.[42] Building on the work of cognitive scientist Peter Gärdenfors, one can ask, What is the geometry of curious thought? And how does it relate to one's own conceptual space, whether ascetic or poetic?[43]

The geometry of the network that we build may change as we watch others practice curiosity. Humans can learn implicitly—without any awareness that they are learning or of what they are learning—simply from visual or auditory observation—that is, by seeing or hearing another's curious acts. We can also learn implicitly from reading another's written work, where curious acts caper in silent symbols across a page. In cognitive science, a commonly studied form of implicit learning is statistical learning, whereby we acquire knowledge about which events happen in our environment, how frequently those events occur, and how often a given event transitions to some other event. In the brain, the neural computations that support this type of learning can facilitate the encoding of pairwise relationships between objects or concepts as well as the encoding of higher-order relational patterns (networks) between them, as we have seen. In learning the practice of curiosity from others, this human capacity could manifest in acquiring knowledge about the types of ideas that others search for, how they connect them, and, over time, how these small steps lead to the growth of knowledge networks.[44] As a complement to these pathways for implicit learning, humans can grow and change through explicit learning, which involves conscious observation, understanding, and memorization of content. We often engage in explicit learning when we realize that we will be rewarded—for example, by a good test score in class or a kind word from a friend. Here our curious behaviors might be shaped by our experiences of a parent, teacher, mentor, or companion referring to our nature as "curious," or our most recent ideas as "curious," thus motivating us to maintain that nature or acquire as well as share similar ideas. What are the places in which we learn the practice of curiosity from others, whether implicitly or explicitly? In a typical classroom, does one learn what category of nodes to look for, types of edges to draw, and sort of networks to build? Does a teacher or professor impart some knowledge about the practice of curiosity, as a by-product of demonstrating their own? Does a mentor transfer a mode of knowledge network building in one-on-one interactions? Does a friend grasp our hand and draw us into the landscape, up the mountain, and through the forest, such that we run together hand in hand along the byways and holloways?

*   *   *

Equipped with this flexible brain, this adaptive mind that learns from others and from our world, we can ask, What are the structures of knowledge that we build? Does knowledge, like the brain, combine flexibility within constraints? Are some parts more rigid than others? Are some parts more flexible than others? Why? Consider an architect constructing the scaffolding of a new building, an engineer constructing the skeleton of a robotic system, or natural selection evolving the mechanics of a living animal.[45] The stability, longevity, and rigidity of such structures is made possible by the geometry of their linkages, as Scottish physicist James Clerk Maxwell pointed out in the mid-1800s.[46] A triangle made from steel bars connected by hinges is rigid because pressure on any bar does not change the triangle's shape. Conversely, a square made from the same materials is flexible because pressure applied to any bar will collapse the square into a rhombus. By combining rigid sections with nonrigid ones, larger structures can be built that marry flexibility and constraints. A structure that bends in the wind, without cracking. The marriage of flexibility and constraints can also be fine-tuned to support a single kinematics. A structure that walks in the wind, without dabbing. It was precisely this fact that Dutch sculptor Theo Jansen used in constructing the *Strandbeests*, a Dutch term that roughly translates to "beach beasts." The power of the ocean wind sweeping along the sands of the shore propels these animal-like mechanical sculptures to walk by the waves as if brought to life, the flexible sections being coupled with the rigid sections in a way that perfectly guides the Strandbeest's gait.

Are these intuitions regarding the relations between structure and movement sequestered to the mechanics of matter alone? Or might they prove to be a useful scaffolding for our understanding of the mechanics of knowledge? As we have seen, knowledge can be usefully modeled as a network of concepts and their relations.[47] Whether well-known or as yet undiscovered, some of those relations are fixed and immutable (such as the relations among whole numbers), while others are flexible in the sense that they can be altered by context or the addition, deletion, or transmutation of concepts around them.[48] As an example of the latter, consider this passage from Arthur Benson's biography of poet and artist Dante Gabriel Rossetti:

> We find such lines as "the unfettered irreversible goal," "Sleepless with cold commemorative eyes." Note such textures as—
>
> > "Oh! what is this that knows the road I came,
> > The flame turned cloud, the cloud returned to flame, The lifted shifted
> > steeps and all the way?—
> > That draws round me at last this wind-warm space, And in regenerate
> > rapture turns my face
> > Upon the devious coverts of dismay?"
>
> > Or
>
> > "Ah! who shall dare to search through what sad maze Thenceforth their
> > incommunicable ways
> > Follow the desultory feed of death!"
>
> It will be observed in these last quotations there is a certain slight shifting of the usual meanings of words like commemorative, regenerate, and

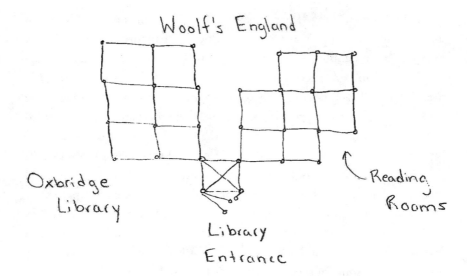

Woolf's England

Oxbridge
Library

Reading
Rooms

Library
Entrance

--- A conformational change in mind ---

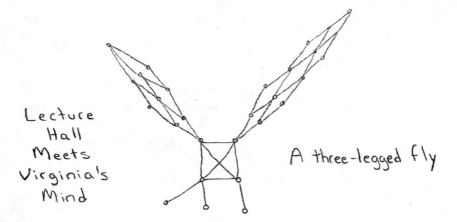

Lecture
Hall
Meets
Virginia's
Mind

A three-legged fly

Figure 7.1

Conformational change. A network could depict an Oxbridge library as it existed in the era of Virginia Woolf. Squares indicate reading rooms, and the additional accoutrements of the library entrance indicate the various doors, security, and so on. Now perhaps my mind moves from the Oxbridge library. But it doesn't just move; it undergoes a conformational change. While rigid triangles must stay fixed, nonrigid squares can flexibly transmute. My mind now sees a three-legged fly in the lecture hall where Woolf heard about the French Revolution. The network is the same (library; fly . . . same number of nodes connected by the same pattern of edges), but the meaning is different (one is an architecture, and the other is a being). This is a conformational change in mind.

incommunicable, some slight nuance added to them which is not found in ordinary speech. This preciosity has a charm of its own, and upon this handling of language, this delicate straining of the use of words, depends much of the pleasure derivable from the work of masters of elaborate style.[49]

Here Benson notes the tendency for Rosetti—a master of language—to stretch, strain, shift, and add nuance to the meaning of words by their placement within a passage. Visual and performing arts similarly innovate within constraints; they recognize the structure and expectations common to most people, and yet seek ways to shift the structure or challenge the expectations just the right amount so that the shift can be appreciated without breaking the structure entirely. And it is from this handling of sound or color, this delicate straining of the use of harmonies or shapes, within the constraints of a given period, that we derive such pleasure. We feel the networks in our mind bend in the wind and walk on the sand.

But network bending is not only the work of artists. It is the work of every human. As we build our knowledge networks through curious practice, do we tether words lightly or even perhaps leave some relatively untethered to allow for their movement in the winds of our future experience? Do we combine sections of rigid (triangular) conceptual relations with sections of flexible (squarish) conceptual relations, allowing our minds to walk or—even more—to dance?[50] When the shape of a mechanical network morphs, like the shapes of the Strandbeests as they walk, we say that the network has undergone a conformational change.[51] Such a change allows proteins in biology to have complex functions, engineered materials to have immense capabilities, and even sculptures to have beauty. Across biology, materials physics, and art, the capacity to undergo a conformational change provides flexible, actuatable function. Could knowledge, whether in a single person's head or in the collective wisdom of the human species, undergo conformational change depending on the precise placement of relational triangles and relational squares? Is it by this shape-morphing process that we gain insight—the "Aha!" moment as two seemingly distant concepts come close enough in our conceptual space to touch? Is it by this shape-morphing process that we bend our beliefs as we educate ourselves about the evidence for pervasive implicit bias along the dimensions of sex, gender, sexual orientation, race, ethnicity, class, and ability? Is it by this shape-morphing process that we mold our closest-held values as we see the benefits of compassion and the devastation of indifference? If so, we seek a kind of curiosity that allows us to build a network of knowledge that blends the flexible and rigid, the squares and triangles, to support continual conformational change. In our minds, upon a landscape that responds to the winds of others and rains of experience, we hope to dance.

8 REIMAGINING EDUCATION

We are pressed into lines, just as lines are the accumulation of such moments of pressure.
—Sara Ahmed, *Queer Phenomenology* (2006)

Among Indigenous peoples of the Great Lakes region and across Anishinaabe culture, the story of curiosity is often told through the sweet secret of maple syrup. According to Mississauga Nishnaabeg writer Leanne Betasamosake Simpson's retelling, a child (Binoojiinh) was wandering out in the woods, collecting firewood—snapping skinny branches and cracking thick ones.[1] We can imagine them curiously observing the comings and goings of things underfoot and things overhead. Moss colonies, as they break down tree and stone into the fertile earth of future forests. Mushrooms, as they draw life from decay, working in tandem with water and wind. The eagle, as it glides by and then pivots to catch a new current. And the thatch of tree canopy, naked tendrils of bough and twig, with leaf buds waiting to burst. Binoojiinh pauses to rest beneath a maple tree. What happened next is unclear. Did they tilt their head up and notice the flurry of energy, or see a bit of bark waft down from above, or hear tiny teeth gnawing the overstory? Did they feel the smallest of tremors in the bark below? Or perhaps Binoojiinh needed no delicate sense of observation, no unusual practice of attention, because two squirrels chasing each other went blundering by and up the trunk. However it happened, Binoojiinh noticed a squirrel (Ajidamoo) scampering about on a limb, gnawing at the bark to sip the sap. Binoojiinh wondered how bark tastes.

Scratching the surface and pressing their lips to the tree, they soon found out. Quickly constructing a makeshift cedar shunt, Binoojiinh collected the sap in a birchbark vessel and then carefully carried it home, leaving their pile of firewood by the tree-side. Reaching their dwelling, they were greeted with excitement and the sap was forthrightly incorporated into the family meal. The residue in the pot that night was thick. The next day, Binoojiinh returned to the tree with family and friends in tow. Syruping became a task they undertook together with themselves and with the forest, never taking too much from any one tree. As such, knowledge grew alongside community practice as well as Binoojiinh's own sense of their capacities to engage with and learn from the world around them.

In Anishinaabe culture, children are enjoined to participate in community practice and knowledge building by finding, using, and celebrating their own inner gifts. What brings Binoojiinh bliss? Do they prefer to experiment alone? Do they learn best with their hands and mouth, fingers and tongue? Are they most curious when at rest or distracted from the task at hand? While some children attend more closely to things underfoot or to those overhead, does Binoojiinh attend best to things in between? After all, the squirrel is a creature that moves easily from turf to canopy in a single scurry. And tree sap, too, travels from the roots to the buds in one season and from the leaves to the roots in another. The gifts of earth and sky and those in between matter, as do the gifts of each dweller in and between worlds. And it is these gifts that are brought to life in community with other humans and the earth. Note that Binoojiinh's newfound knowledge, built in synchrony with the squirrel, moreover, was welcomed. It might have been otherwise. One can imagine a situation in which a particular people prize knowledge of moss and mushroom, eagle and wind, but not squirrel. In this scenario, squirrel knowledge is disparaged, considered to be not serious knowledge (too interdisciplinary? too undisciplined?), so to speak. For them, true knowledge is of things underfoot or overhead, not in between. But of course, Binoojiinh's community is different, and welcomes whatever the knowledge is and from wherever it has come, as long as it is rooted in respect for Binoojiinh's own gifts and those of the land in which they find themselves.

Long before Binoojiinh's discovery, as Robin Wall Kimmerer tells it, Indigenous peoples forgot these values of relational curiosity, activated via individual gifts and community efforts in concert with the teachings of the land.[2] In this fabled past, it was no longer a question of who was better suited to gardening or hunting, safeguarding or innovating. Everything lay in disarray as entire communities parked their bodies beneath the maple trees, mouths agape, drinking the sweet syrup without care or context. They paid no regard to the trees, paid no respect to the value of reciprocity, and paid no mind to balance, whether the balance of their own nourishment or that of the forest. To right that imbalance, a spirit leader diluted the maples until there was only a hint left of their sweetness: sap. When read alongside Simpson's retelling of Binoojiinh's story, we wonder if this isn't a story of curiosity gone wrong. Mouths agape: is this an egoistic form of knowledge acquisition? Is this innovation for innovation's sake? Is this curiosity hitched to capital? Is this curiosity-about, rather than curiosity-with? The two stories alongside one another seem to pose the critical question of how to foster learning environments that are characterized not by a sea of individuals consuming information (guzzling from their iPhones or researching without regard for ethics) but rather a network of young and old folks who, in discovering new edges, build ever-richer webs of existential and epistemic connection.

In this chapter, we grapple not only with the *conditions* that make learning possible but also with the *kind* of learning we want to make possible. Following Kimmerer's invitation to "pair" Indigenous wisdom and Western science, we herein stage certain resonances between Binoojiinh's story, network science, and other critical pedagogies, all while taking to heart the resurgent Indigenous claim that learning necessarily exceeds mainstream educational institutions.[3] We do so with the understanding that fundamental to both Indigenous ontologies and a network approach is this foundational insight: that beneath the seemingly random smattering of things, there lies a structure—indeed multiple structures—of connection. Given the existing dynamics of these complex relational systems, how might we best learn and educate, lead and follow? To answer, we reconsider the diversity of neural, social, and knowledge networks. A curiosity rooted in the gifts and social worlds of each learner, we argue, necessarily results in knowledge networks that grow in unexpected, often unruly directions, with divergent—and defiant—architectures. Exploring the practical implications of reconceptualizing curiosity as edgework, we posit that the space of learning must necessarily change. And this change must come about not merely for any number of proximal goals but also for the sake of the many squirrels—and squirrel knowledges—yet to be heard.

FROM PARTICLES TO PATTERNS

How do we practice curiosity? How do we investigate, explore, and discover? And how do we do it together? As the human species, we pursue our questions in nature and in the city, in the lab and on the dance floor, through spacecrafts and on walks, protesting and in prayer. Of all the spaces that facilitate collective human curiosity, however, perhaps the most paradigmatic are places of education. These are spaces in which we gather to commemorate and to create tracks of knowledge. By turns, we lead and follow one another along the furrows, crisscrossing the world of things known and yet to be known. Nevertheless, as academic studies and personal testimonials repeatedly confirm, educational institutions today are consistently threatened with the loss of curiosity, a lagging of wonder, and a dearth of delight in discovery.[4] Too often, compliance and complacency, bureaucratic constraints and mundane deliverables compromise the environments of creative freedom necessary for the pursuit of knowledge. And far too often, social inequalities, ableism, and disciplinary silos build walls between learners, their potentialities, and their communities. It is for these reasons that learners and educators alike repeatedly ask, By what pathways can the very institutions of curiosity remain curious? How can we imagine learning differently, more equitably? How can these spaces work in concert with the diversity and relational fabric of knowers as well as the connective tissue of knowledge itself?

One of the fundamental problems with mainstream educational practice is that it still largely targets the single student, with a standard mind, through a set curriculum. The fault in our stars is that they fall to dust in our hands when extracted from the constellations and galaxies that lend them meaning and function. Too often, the complexity of the learning environment is lost in favor of the implicitly presumed capacities and needs of the standard student. The classroom has historically been largely homogeneous, artificially crafted through the segregation of women, students of color, and students with disabilities. As contemporary classrooms have become increasingly diverse, the specter of homogeneity remains palpable in standardized curricula and evaluations. Just as sociocultural differences are often either sidelined or stereotyped, so learning differences are often either ignored or reinforced. While students with disabilities are shuttled into different programs, even neurotypical students who learn in a vast majority of ways are disadvantaged by mainstream pedagogies that continue to presume largely homogeneous brain network structures. Developing a flexible approach *across* the complexities of language, cultural heritage, social behavior, intellectual capacities, and affinities is therefore necessary in order to make the classroom one in which all students can flourish. Doing so, moreover, means forgoing stiff subject divisions and disciplinary silos. The act of facilitating the living connections in learners' minds and communities, as well as between topics and methods of inquiry, trades an isolated education for deep network learning.

By foregrounding a network perspective, as specifically applied to brain structures, social groups, and systems of knowledge, we can envision curiosity as precisely not a discrete capacity of individuals but rather as a growth principle for interconnections in and among various architectures. While curiosity is easily curtailed in ableist settings, curiosity can flourish as a pluriform practice flexible across different brain structures and capacities. Likewise, while curiosity is easily squelched in segregated settings or spaces saturated in stereotype threat, curiosity can also serve as a bonding agent across groups, lifting people from the confines of their original contexts to appreciatively engage across difference. Ideally, learning facilitators would ask not who is curious, but how is each person curious? What is the style of their exploration and the timbre of their queries? Such an approach opens up the possibility for people who think differently to, in a real sense, lead the way, cultivating fresh perspectives, trenchant critiques, and signals heretofore mistaken for noise. Finally, while curiosity can succumb to the blight of ritualized methods and brittle distinctions, curiosity at its best is discipline independent, functioning across and through and outside of existing disciplinary purviews and procedures. Thus, appreciating the networks of knowledge, social bonds, and neural substrates not only preconditions a reconfiguration of education but also reframes what curiosity is, and how it can be facilitated in more vibrant and equitable ways.[5]

This chapter, then, is an experiment in imagination. It imagines diversifying curiosity. It imagines democratizing curiosity. And it imagines de-disciplining curiosity. Committed to reducing the social inequities and softening the coarse disciplinary boundaries that subtend the academy, we aim to facilitate freedom for *everyone's* curiosity to pursue seemingly incommensurable ideas and applications. We sketch out the paths of possibility that beckon us into a more inclusive, transdisciplinary pedagogy. And we wager that as we understand curious minds better, we can better support more curious futures.

DOTS OUT OF LINE: ON NEUROATYPICAL CURIOSITY

How is your brain built? How are its neural networks structured? Which neural tracks prefer to fire with which and in what context? And how big is the fire? Which like to dance and what is their style? How do two and two get interlaced, and hang together or float away? And how is all of this different from or similar to the person next to you in a space of learning? Some curiosities look like scampering squirrels or eagles in flight, while others get overlooked like moss underfoot. But curiosity need not be flashy or even strong. Much like moss, it needs only a bit of porous surface in which to sink its rhizoids and the barest of breezes toward which to unfurl. And yet without moss, the world of squirrel and eagle would be impossible. Secreting acids into stone, moss makes way for woodlands. A few cracks and a light patience—the slow burn of critique and a little wonder—give way to life.

Born in 1992 and diagnosed with autism of a "severe" and "nonverbal" type, Naoki Higashida is the author of dozens of poems, short stories, and nonfiction books. He attended a number of educational institutions in the course of his youth, including a neurotypically centered primary school, a special needs junior high, and a distance learning high school. Observing the many ways in which people with disabilities are fast-tracked out of normal life and shuttled into special needs circles (and low-wage jobs, if they are lucky), Higashida wanted to make his own choice. After "questioning things" for himself, and identifying his gifts and hopes, he decided to become a writer.[6] While in school, he had become increasingly frustrated with the steady infantilization and "schoolmaster-type instruction" that denied his creativity and squashed his curiosity. Indeed, scientists and educators alike repeatedly characterize people with autism as lacking any measurable—and therefore meaningful—curiosity.[7] Resisting this narrative in his writing and activism, Higashida repeatedly asserts his own curiosity: "I'm always hungry to learn," "hungry for knowledge," he writes; "I want to grow up learning a million things!"[8] Other autistic people, he says, are much like him in this respect, "constantly challenging and asking questions of themselves." This should be no real surprise, he writes in a poem called "Curiosity," given that curiosity is fundamentally human; it "is why we carry on."[9]

Of course, because the neural structures in Higashida's mind are atypical, his curiosity functions—and manifests itself—differently. For him, there is a certain absorption that characterizes his interaction with the world. Time and space are nebulous, as are the boundaries of his body and the contents of his mind. Exploration therefore takes on a different timbre and perceptual perturbation a different hue. In greeting the world, Higashida remarks, "details jump straight out," while objects themselves linger in the background; as such, there is "a world of difference between seeing a thing and knowing what it is."[10] Coming to know, to recognize, and to engage with things requires shuttling between the present experience and a series of informative memories. Higashida's memories, however, are not stored in carefully indexed and chronologically arranged slots. They are a jumble of "dots" that float about "scattershot."[11] Sometimes he can patiently "fish" for the right one, while at other times it's like reaching for a star hung "hundreds of millions of light-years away." His facility in jumping in and out of the sea—or in and out of the galaxy—is a bit of magic, and words help make that magic happen. Communicating via keyboard, Higashida is transformed into a "dolphin," able to swim and flip in the vast ocean of language, and chatter away with his companions. Through writing, he weaves together, with passion as much as patience, the "connections" between his knowledge and his experience, his own mind and that of others.[12]

If neurotypical memories—and all the things that have been learned—are structured more like an index, what is the architecture of that index? What is its shape, or conversely, how does it lend shape to what is? By what logic are its folds fashioned? In contrast, what happens when one's knowledge looks more like a school of fish or a galaxy of stars? And what about dots and dolphins? Shape and movement in this case have an entirely different physics. How does one walk that graph? What are the steps by which curiosity moves and how are its paths interwoven? First and foremost, Higashida's work testifies to his own curiosity, flush with obsessions and fixations, interests and fascinations. He asserts that his curiosity takes different shapes and functions by a different logic, and that this logic can be beautiful all on its own. His work secondarily enjoins neurotypical people to treat him and other autistic people as "neither a princeling nor a spare part" but rather with compassion.[13] Our capacities and curiosities—*our futures*, he insists—are "connected" with one another.[14] What does that look like in a space—indeed a world—of learning, where fish curiosities, star curiosities, and index card curiosities sit side by side? How do we engage the world alongside each other and even with one another? What are the markers of a learning environment in which fish and filing systems can equally flourish?

Scholars who study curiosity have paid remarkably little attention to neuroatypical learners, let alone neuroatypical curiosity. Attention to diverse neural networks, however, requires precisely this. The network approach equips us to think much more richly about the validity of different modes of curiosity. Consider a room full of students with a range of learning differences, from dyslexia and autism, to anxiety, depression, and bipolar disorder, or again from hyperactivity and obsessive-compulsive tendencies all the way to Gardner's multiple intelligences. Each learner has their own architectures of learning.[15] How wide is that slate of curiosity signatures?[16] So much more could be done to understand and celebrate the often nonstandard ways in which curiosity is expressed.[17] New means of recognizing nonverbal, sensory, and affinity interests are needed in order to honor the curiosity in people who have long been thought incurious.[18] People largely treated as zero-dimensional dots may themselves be masterful dot conductors. Such a project demands a new series of questions. What makes something questionable for this person or intriguing for that one? How are questions differently asked and formulated, generated and shaped? Ultimately, what *sorts* of information does each seek, in what distinct *manners*, and resulting in what *actions*?[19] How is the practice of curiosity variable even within the individual and across timescales?[20] How is each learner's style of curiosity correlated with their neural structures and sociocultural environments? And how is that capaciousness reflected in the curriculum?

Such work necessarily begins with a commitment foundational to disability communities and disability studies: that disability and neuroatypicality are normal variations in bodyminds and that, as such, people with those variations are equally curious cultivators in the field of knowledge. Network neuroscience provides exceptionally useful tools for illuminating the breadth of neural structures and functions across a variety of typical and atypical learners. Its research into attentional capacities, memory processes, brain plasticity, individual differences, and genetic and epigenetic factors has the capacity, moreover, to clarify not only the architecture of the learning organ (i.e., the brain) but also the architectures of learnability itself.[21] All of this equips us to ask the question, What makes for optimal learning in new ways and *optimal learning for whom*? From the vantage point of disability theorists, however, neuroatypical learners and their appropriate pedagogies are not meant to be added to the mix but rather to transform the learning process for all.[22] Universal design (originally an architectural concept) puts multidimensional and multisensory learning opportunities at the center of education in order to facilitate access for the broadest swath of minds.[23] Expressly citing differences in brain networks, the universal design framework multiplies educative avenues, whether for curious expression and engagement or for instruction and assessment.[24] Designing for everyone instead of for most, learners and educators alike can enjoy a greater range of knowledge-building practices, atypical or otherwise.

Diversifying curiosity involves enlarging the range of recognized inquisitive practices and affects. From whatever walk of life students hail, there is a vast range of ways in which each moves through social and conceptual space, and a real variety of walks they take on knowledge networks in any given day. They run, skip, and limp, trudge, wheel, and drag; sometimes they circle, other times they strike out yonder. The network approach to curiosity helps educators to facilitate epistemic growth in a landscape enriched by the "biological exuberance" of structural and functional difference—the whole ecology of the forest.[25]

SOUL POWER: ON CURIOSITY AT THE MARGINS

How is knowledge built—and how is curiosity practiced—within and across different social networks? How does the shape of learning change across geography and history? How are cultural values and habits embedded in what gets asked, when and where it gets asked, and how? Much like forest biomes, within which certain trees, plants, and animals wax and wane in the tangled web of one another's company, ecosystems of knowledge differ and decay by turns. Sometimes vast swaths of knowers and knowledges are razed to the ground, much like clear-cuts. In their wake, one sort of knower and one sort of knowledge is planted, reminiscent of tree plantations. Epistemic monocultures, however, reduce growth while increasing susceptibility to blight, soil erosion, and nutrient depletion. By contrast, in epistemic polycultures, where a range of knowers and knowledges from different social groups are cultivated beside one another, growth and resilience across groups is enhanced. How might different knowledge communities be celebrated? How might inequities between knowledge communities be addressed? And how might the traffic between worlds of knowing be cultivated so as to frustrate eliminative and extractive epistemologies, and instead facilitate respect and accountability?

In the summer of 1964, the Freedom Schools arose as part of a larger effort to offer free, liberatory education especially to Black high school students in the US South. It had become patently clear to civil rights activists that educating the youth, "whether white or black, rich or poor," was essential to the movement.[26] While the curriculum included basic math and language education, its primary contribution was a deeply existential and political experience. Freedom Schools aimed to provide an honest, vibrant context in which students could ask about the nature of themselves and their world, and how they fit into it or were alienated from it.[27] This is in stark contrast to the sort of learning experience they had the rest of the year. In the public schools, the curriculum reflected white Euro-American history and values, leaving the African diaspora submerged in a well of silence. There were teachers rather than leaders. And they taught material instead of facilitating relationships. Taught to be objective and to process material via information dumping, students learned content abstracted from their daily lives. "Are people 'educated,'" though, "if they know everything *except* their own lives?" one of the organizers asked.[28] In their regular schools, Black students were typically not asked to question things themselves. And if they did try to investigate racial politics, they were often suspended or expelled. Each student was taught to learn "things" as an isolated learner, abstracted from their communities and alienated from their world.

If public schools channeled and policed curiosity in a way that served to maintain white supremacy in the United States and a segregated status quo, the Freedom Schools facilitated curiosity differently. The schools opened curiosity back up, and redirected it to the lived realities of all students and especially Black students, for it is in the very phenomena of existence that politics dwells. Repeatedly, the Freedom Schools' facilitators and curriculum identified "questioning" as the heart of their effort—to question everything and to question it earnestly. To ask the "real questions" that invite Black students to listen to themselves, others, and their world in a way that they never have.[29] Organizers granted that this work is painful, but it is also the sole source of empowerment. It is "the vital tool" to becoming "active agents" of social change along with building racial and class coalition. The Freedom Schools' curriculum itself was half questions, with a centerpiece of its "citizenship education" almost entirely constituted by questions. These questions—which covered issues of education, employment, housing, medical care, policing, voting rights, stereotypes, fears, power structures, and powers of resistance (i.e., truth power and soul power)—were not simply a collection of questions, organizers explained, but instead a web in which each question "springs from a question" and "leads on to another question." Through this web of questions, an activated and activist social network can be built, one founded—for them—on true democratic practice.[30]

The Mississippi Freedom Schools posed several fundamental questions to guide all of their inquiries. Among them, two are paramount: "What does the majority culture have that we want?" and "What do we have that we want to keep?" Black high school students, in particular, were repeatedly invited to interrogate the effects of social inequity across knowledge networks, and then to insist, for example, that their schools be stocked appropriately and that they themselves be taken seriously. But students were also repeatedly invited to think about existing practices of curiosity and knowledge building within Black communities of the US South. What habits were important to preserve and what forms of meaning making were critical to recover? Ultimately, the Freedom Schools were a testament not only to the ways in which institutionalized curiosity in schools can serve the status quo but also to the power of resistant curiosity, built specifically in Black communities, in poor communities, and at the political margins, to reimagine and reconfigure the social world—or rather *social worlds*. While atomized curiosity may solidify accepted knowledge, relational curiosity forces facilitators and learners alike to critically reengineer knowledge built in and across different social networks, and the pathways of travel between them, as well as to address the virulent inequities so long inherited.

What matters here is both who learns together and what they learn. What social worlds, what histories, and what queries are germane to the space of learning? For underrepresented students, there is a clear cost to learning in settler spaces, white spaces, ableist spaces, patriarchal spaces, and straight/cisgender-dominated spaces. With obdurate homophily and recalcitrant gatekeepers, especially among the more privileged classes, and the timeless correlation of a social network's central positions with socio-economic status, marginalized actors are consistently limited in their ability to shape the educational space as well as their own futures. While robust social networks are highly correlated with learning performance, students from marginalized groups—including women, students of color, as well as first-generation, disabled, and/or LGBTQ students—traditionally lack those networks.[31] This differential distribution of connectedness compromises what students can be empowered to ask of themselves and of their world. But as the Freedom Schools attest, while educational institutions must be reimagined, they are not the only source of hope. Long histories of refusal have facilitated radical learning in unofficial venues and at unauthorized times. Tracking these illicit streams of social belonging and social inquiry illuminates not only the necessity but also the possibility of extrajudicial learning spaces—and the power of their curiosity—to fundamentally disrupt social hierarchies.[32]

Curiosity is socially curated as much as it is socially cultivated. Even if some students are most curious in solitude, they query within a larger matrix. What would it take to generate greater equity in education? Critical pedagogues suggest that acquisitional curiosity actually contributes to the homogenization of the learning environment. Theoretically, then, curiosity as edgework—that is, as fundamentally relational—might fruitfully disrupt the learning environment with the promises and improprieties of epistemic divergence. Educator Paulo Freire, for example, "refused" the banking model of education whereby teachers impart standard information to standard students. Instead, he endorsed a pedagogy of "communitarian praxis" whereby, for him, Global South learners think *together* and from within the space of their experience, and thereby become "makers and dreamers of history."[33] Democratizing curiosity necessitates rebalancing the hubs and spokes that have for centuries kept certain knowledge systems and knowers peripheral to, if not isolated from, reigning structures of knowledge production. It involves extending and reorienting the graph, and more conscientiously creating the pathways across it. Feminist author and activist bell hooks, in her consideration of sexism and anti-Black racism in US schools, characterizes true education as a "practice of freedom," wherein learning itself is a "revolution." Such learning is predicated on redrawing the webs of our relationships to foreground "our interest in one another, in hearing one another's voices, [and] in recognizing one another's presence," across gender, race and ethnicity, class, ability, and sexual orientation.[34] It involves daring to elevate long histories of inquiry from marginalized corners of the world—and from all the corners of human experience. And it involves developing concrete practical habits of listening between and across different communities of inquiry.

Social networks are inherently shared, but they are not inherently equitable. Developing healthy social networks in and around learning communities, and correlatively robust networks of curiosity, requires reconfiguring existing architectures. This involves new (or sometimes very old) modes of study, which are themselves correlated with new (or sometimes very old) modes of self making and world making.[35] These modes of study in and outside traditional educational spaces may include storytelling, political organizing, artistic creation, jamming (music and machinery), meditation, land-based practice, access mapping, and so on. When social differences and inequities are foregrounded, democratic education necessarily moves beyond reforming the teacher-to-student dyad to deeply redistributing networks of inquiry and of inquisitiveness. In this effort, the social nature of curiosity, whereby learners grow to mimic the topics and methods of inquiry modeled by their mentors and friends, is no mere fact but instead a weighty responsibility.[36]

NO SUCH THING: ON CURIOSITY DE-DISCIPLINED

How does curriculum get decided? How are sciences solidified and canons of knowledge crafted? How are epistemic discourses formed and fashioned? And by what techniques are they transformed, whether in sanctioned settings or otherwise? People often treat knowledge as if it lives in silos, and is harvested only from this field and not that one. But knowledge grows in the wild and the crosscurrents of curiosity run underground like mycorrhizal networks, shuttling food for thought between knowing organisms and knowledge systems. The vast majority of plants exchange resources and communicate with one another through the miles of threadlike hyphae underfoot. Fungi give the lie to any illusion of a monadic forest. So, too, are the stark distinctions between knowledge enterprises ultimately foolish. How might those interconnections and those intimacies surface in and transform the space of learning?

Leonardo da Vinci is repeatedly celebrated as one of "the most curious minds to ever have existed."[37] The proof, it is said, lies in the very range of his mind, moving raucously from math, science, and technology, to art, music, and physiognomy. What is distinctive about da Vinci, however, is not only the breadth of his interests but also the way those interests straddle time. At one moment he is buried in anatomical intricacies right before his eyes (e.g., bears and frogs), while at others he is blithely blueprinting inventions as yet impossible to comprehend (e.g., hydraulic pumps and helicopters). Da Vinci could cradle the now and catapult into the future in a single flourish. It is perhaps for this reason that psychoanalyst Sigmund Freud characterizes da Vinci's curiosity as "a single impulse very forcibly developed."[38] And it was unstoppable. Da Vinci would go about his days compulsively note taking and sketching, his drafts as meticulously crafted as they were provisional in nature, as if Curiosity were an exacting and yet changeable muse. The paradigmatic Renaissance man, da Vinci was nevertheless a man without letters, being surprisingly unable to read Greek and barely able to pick out Latin. His was therefore a self-taught, self-driven erudition, fueled by a curiosity untamed by the expectations of scholarly acumen of his time.

Not everyone is celebrated for such things. Interdisciplinarity and antidisciplinarity—the capacity to think across and beyond established frames of knowledge—can be heavily disparaged depending on who you are and where your curiosity takes you. Consider a brief series of testaments. Kimmerer is an Indigenous botanist. During her first-year advising session, she spoke of her enthrallment with the aesthetic harmony of asters and goldenrod; her college adviser quickly replied, "That is not science. That is not at all the sort of thing with which botanists concern themselves."[39] Alison Kafer is a disability theorist. During graduate school, she proposed a paper on then cutting-edge cultural approaches to disability;

her professor rejected it as "insufficiently academic."[40] Gloria Anzaldúa is a third-wave feminist icon. During her PhD in comparative literature, she purposed to study Chicana feminism; her adviser said that "Chicana literature was not a legitimate discipline, . . . it didn't exist," and recommended she stick with "red, white, and blue" disciplines and ways of knowing.[41] Shay Welch is a professor of Native American philosophy. She applied for funding from the National Endowment for the Humanities to support a project on Indigenous epistemologies and was told by a reviewer, "There is no such thing as Native American Philosophy. Native peoples did not produce anything sophisticated or systematic enough to constitute proper Philosophy."[42] Now cornerstones of their subfields, fields, or founders of new fields, these women, following the beck and call of their curiosities, rooted in themselves and their communities, were nevertheless told "no." Thankfully, they didn't listen.

While curiosity at the edge is celebrated in some contexts, it is starkly policed in others. What is it like to be one of these *remolinos*, these edge-walkers? What would it be like to learn from these many lives, their communities and their alternative histories? When curiosity is rooted in individual gifts and community engagement, knowledge will move in unexpected directions and take on heretofore-unimaginable shapes. Such movement can be frustrated by bias, prejudice, and other forms of small-mindedness, but also by too pure an attachment to knowledge structures *as they are* instead of to the architectures that might be.

If knowledge is a network that grows and changes by adding and subtracting nodes, weighting and hollowing out edges, then what is curiosity? Of course, curiosity can function within severely restricted contexts, building new knowledge in only the most sanctioned of directions, via approved means and methods.[43] The redemptive promise of curiosity, however, lies in its transgressive and dynamic potential. It leaps from hub to hub, randomly walks the grid surface, and lightly skips from core positions to peripheries. Reckoning with knowledge as a network, and specifically with the challenge of epistemic justice between modular networks, then, entails de-disciplining curiosity.[44] For us, "de-disciplining curiosity" does not mean undisciplining curiosity in any simple sense. It does not mean forswearing all methods and canons but rather working within, yet also *across* and *beyond*, those methods and canons in ways that productively resituate, reorient, and in some cases undo them. It is, after all, through disciplinary cracks both within and at the edge that new things can grow. In this sense, de-disciplining carries echoes of the interdisciplinary, transdisciplinary, and even antidisciplinary. Like interdisciplinary work, de-disciplined curiosity might combine discrete concepts, theories, or methods so as to pursue interconnected ways of knowing, but also subject field-specific boundaries to critical questioning.[45] Like transdisciplinary work, de-disciplined curiosity might crisscross norms and positions so as to recontour the terrain. And it might do so from an errant, "risk-taking" consciousness often characteristic of "rekindlers of hope."[46] Finally, like antidisciplinary work, de-disciplined curiosity might have no settled relation to established practices but instead court the flexibility necessary to nonutilitarian and emancipatory inquiry.[47] Insofar as disciplines and curricular subjects are segregated and siloed, however loudly academic and educational spaces claim otherwise, a flourishing ecosystem of knowledge must necessarily engage more dynamic characteristics and contributions of curiosity.

What does this mean for education? And for vibrant education specifically? Students and teachers alike each enter the classroom or study group with unique knowledge networks already formed and fashioned. Those knowledge networks are informed by the collective knowledge held by the geoculturally defined subgroups of humans to which the knower somehow belongs. That collective knowledge network, likewise, is informed by extant human knowledge across time and space. The student and teacher enter the learning environment in order to grow as knowers. But in which direction will their knowledge networks extend? And what unfamiliar knowledges will they grow to honor? Because collective knowledge regimes are subject to contingent constraints, whereby certain knowers and knowledges are, by turns, restricted to certain quadrants and excluded from others, it is more than likely that student and teacher knowledge network growth will skew likewise. The processes that shape the categories by which we all understand the world are not only built on social bias and historical inequities, as well as economies of epistemic necessity, but also on disciplinary and academic purism.[48] There is vast space left on the grid for outsider theory, for unfashionable ideas by unfashionable people, those excluded from or excluded within the academy.[49] And there are whole worlds of knowledge invisible to one another. Given the relevant small-world networks, the long links between hubs or across grids are far less common than the short, internal ones. Yet they signal the infinite plots of knowledge left to the peripheries of social graphs, if not uninscribed entirely. An active de-disciplining of restrictive knowledge networks would help mainstream educational experiences to be less predicated on epistemologies of ignorance and more engaged in acts of cross-network creativity and appreciation.

To democratizing and diversifying curiosity, then, knowledge network theory contributes an insistence on de-disciplining curiosity. Far from a license to make half-hearted and superficial queries, such a de-disciplining of curiosity reinscribes inquiry within the fragile demands of freedom in a modular world. In educational settings, it reroots learning not only in real-world projects and wholistic pedagogy but also within and across communities of practice through which one might ask as yet unthinkable, inconceivable questions. In doing so, curiosity reinvigorates the gut and grime of the graph, enhancing the mysteries that lie in their interstices and just beyond the edge.

FROM PATTERNS TO PEDAGOGY

How might we then teach and how might we then learn? What pathways and byways suddenly present themselves before our eyes as we attune ourselves to the beautiful complexity of a networked life? And what might be the limitations implicit in even this expansive framework, so inherently sensitized to difference and collectivity? We have argued that, in applying network theory to brain structures, social groups, and systems of knowledge production, we become more equipped to imagine curiosity differently. Far from a discrete capacity more or less present in individual knowers, curiosity is a growth principle for interconnections among people, neural architectures, and bodies of knowledge. As such, curiosity is best diversified, democratized, and de-disciplined. A network science and philosophy of curiosity therefore clears the path for a more inclusive, transdisciplinary pedagogy. It is important to recognize, however, that while such a pedagogy allows for greater individualization in education, that individualization is won only through a greater collectivization of curiosity. It requires first appreciating the ways in which the "we" who cross and crisscross the furrows of knowledge and things to be known represent a vast array of knowers and ways of knowing.

There are a whole host of ways to explore network-infused pedagogies in the future. As we do so, it is important to remember that the social, neural, and epistemic networks in and through which we learn are more diverse than previously acknowledged. But it is equally important to recognize that those networks did not just appear on the scene fully baked in their present formations. What we learn and what we teach, our networks of belonging and sense making, are never ground zero, nor are they ever truly singular. Everything has a history to its reticular growth and a story to its network structure. The recognized forms of learning and recognized types of learners have a history. The ways in which social histories get told and social differences get adjudicated have a history. And the things deemed important to know in order to be educated in this or that field—or indeed to be educated at all—have a history. Part of the learning process involves appreciating that all of these networks come from someplace and are going somewhere. They are embedded in values and investments, factions and debates. What can we learn from those histories and what futures can we create in light of them? To teach and to learn in our contemporary moment is to leap onto a scene of struggle over inquiry, how it has been done, is done, and might yet be done. And it is to suddenly have in our hands the capacity to enter the fray and propose what learning might look like, how knowledge might be acquired, and how curiosity might best be practiced. We get the chance to redraw the shape of our worlds together.

Schools and universities have a responsibility to cultivate a critical awareness of the networks that inform what kinds of questions we and our students do and do not ask in the educational setting. As practitioners of curiosity, we and our students likewise have a responsibility to hold our institutions accountable to that critical awareness. But these places do not exhaust the spaces of learning. We might think of eagles as only ever flying overhead, but they rest on trees and cliff faces, dive between open fields and lakes, and waft on global currents. Learning, too, can hardly be confined to institutions. As Simpson and Kimmerer insist, there is a whole world to be engaged. And this is where Indigenous thought is fundamentally still deeper than the network approach we have offered here. Basic to Indigenous resurgence is not only a right to land but a relationship with land and the networks of that land.[50] To reimagine curiosity only with reference to human brain networks, human social networks, and human knowledge networks is to fail to appreciate the more-than-human world in which curiosity circulates. Indigenous thought deeply challenges mainstream educational systems to account for the environmental alienations and costs that are basic to their function. And it insists on an apprenticeship to the wisdoms of the world around us. This call lends a greater urgency to the recognition that our questions are always coming from somewhere and doing something. The clearer we get about the who, what, where, and why of curiosity, as investigators and learners alike, the better we can appreciate our complicated histories, engage our earth-bound presents, and craft coalitional futures.

*　　*　　*

But back to squirrels. By many accounts, squirrels seem to embody all of curiosity's worst tendencies. Squirrels seem erratic, aided and abetted by their double-jointed heels. And their constant chatter has earned them the mythological reputation of first-class troublemakers. They appear unreliable, catapulting between worlds underfoot and overhead (there and back again), and they are even fabled to have swallowed the sun once or twice. They sip the sap, sure, but more often they eat from their cache of nuts and seeds, which they have accumulated relentlessly in the prewinter months. But we wonder if this isn't much like acquisition at all. For squirrels, theirs is a storehouse of memories, edging around their forests, down into ravines, and round rolling hills. Every bit of nutrition is planted (not stock-piled in an arid cellar or private attic). Planted in such a way that someday it will nourish a squirrel or some other creature, or—a not insignificant number of times—it will germinate and flower. Something of a gardener and ecologist, the squirrel gathers and plants, gathers and plants, building a relational network of possibilities for a still-unimaginable elsewhere and elsewhen. It holds its discoveries with an open hand. Whatever the future of education might be, perhaps this creature can teach us still more about curiosity.

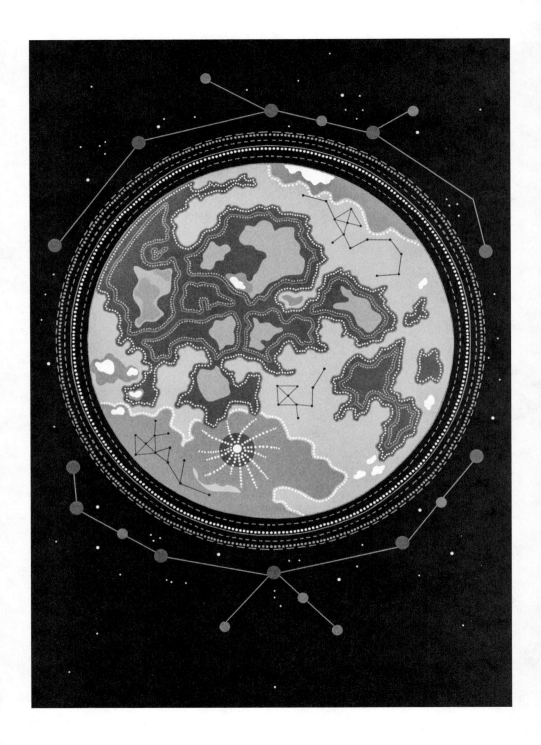

CONCLUSION: ON CURIOUS FUTURES

When we think of curious futures, futures of greater freedom for the vivacity and variability of curiosity, we think not only about what can be built but also about the cracks in those edifices. Cracks birthed in the past. Cracks just opening today. Cracks still waiting to inhabit the future. Curious lives need freedom to break from lines inherited, even from fresh threads of self and society, and from trails that bind us in any number of ways. Curious lives bring with them a bravery to unweave that which is woven, a patient mode of undoing, and a hankering to de-resin the monolith, reveal the fissures, and weaken the stone. Curious lives harbor a tenacity to unknow and unsee, to reexamine and problematize, to make a hole where there wasn't one before, and to imagine *through a crack* into that which is outside of and beyond. The most flexible mode of formation is one that freely unforms, writes and erases and rewrites, cuts and pastes, rivets and rives. "Build a web and then eat it—three times a day," wisely advises our eight-legged friend.

CURIOUSLY CRACKING

It has become something of a truism that curiosity is a drive to fill information gaps. We have proposed instead that curiosity is edgework. It is a process of building connections and relations between things in the quest of coming to know them. Edges cross gaps between things not in order to fill them but in order to traverse the space, to reconnoiter the shapes of solid things and draw new lines. Socrates, a paradigmatically curious fellow, is remembered for pestering people with questions, relentlessly pressing them until they discover one fundamental fact: they are ignorant of the thing they thought they knew. In a sense, Socrates's curiosity was the cracking kind. He hoped we would lose our heads a bit. Rather than filling information gaps, his curiosity made the gap grow ever wider, as if a cavern suddenly yawned from the ground beneath our feet. While we have focused throughout this book on curiosity's building work, we close with this musing on cracks in order to bring to the fore the forces of unbuilding and unknowing that are equally at work in curiosity. Cracks allow edge-work to work; they give it something to do. If, then, curiosity is to have a sound and vibrant future, it must crack.

The word *crack* first appeared on the scene in Old English about one thousand years ago. It lived for several hundred years exclusively as a verb. In its verb life, it had many amazing adventures, leaping from poet's pen to *Bürgermeister*'s harrumph. It marked, more than anything, sound. Sound, of course, moves in waves, but the sound of cracking moves like waves toppling into a cleft or over a cliff. Crack! The blast of a break. Cracking can come in big ways and small (a bit of crackling). There are the cracks of sky and earth from a storm overhead or from deep within. But there is also the sound of old bones in the morning or of young wood on the hearth. And it is not uncommon for the rogue companion or two to find themselves cracking jokes one moment and cracking bottles of mead over each other's heads the next. Three or four hundred years later, the word *crack* was recodified, now as a noun. It became the break that sounded the blast rather than the blast itself. It was the fracture—or fissure—that formed, whether anyone ever heard it or no. And it was the crack that never even had a sound, like the silent crack beneath a door through which air slips quietly, eddying, wafting, with little urgency and no real beginning. A crack, then, much like a cleft in a cliff face, may be from a sharp break or worn away over centuries. It is a certain unsoundness of a thing, with, perhaps, no sound to tell the tale. Fundaments foreclose a certain kind of curiosity, whereas cracks make curiosity possible. Or perhaps it is curiosity that cracks things up. The crack of a question sounds and resounds where certain forces have long presided over a crafted silence, where doubts have been muffled and queries quickly suffocated. Curiosity surges into the break. And curious minds are cracked in the most beautiful and vibrant of ways. They do not fear fractures of mind and of feeling, nor do they avoid ruptures of sense and of fealty. They wait for them, patiently and yet eagerly, as one waits for the newfound fervor of spring. Indeed, curious futures are filled with these fecund cracks in ourselves, in our world, and on our horizon, and we can learn to navigate these cracks by looking first to the natural world, where permanent, induced, or volunteered fissures fuel the circle of life. Much like an ellipsis, a crack is "a point of exit, from the visible toward the invisible."[1] Curiosity makes more things possible by making more things, patiently and precisely, impossible. By breaking with them and cracking them open.

*   *   *

"Here's to cracking, then!" we say, as we settle down to a day spent together doing our favorite things. We begin by cracking open Curiosity's dictionary and we find the following entry:

> (*vb.*) to crack (permanently); to split apart, to fracture; when that within breaks free of that without.

Physical ruptures appear in all kinds of surfaces when the materials under, within, or behind them must pass through, without, or beyond them. In the work of a moment, the surface is roughened and the solid is riven. The transgression of the material's boundary feels much like an adventurous traipsing across the void, and with it comes new life, balance, and liberty. *New life* in the clear-white and translucent-green tones of the embryonic root, or radicle, as it emerges from the seed coat in a slow breach of the water-laden capsule. *Balance* in the fiery brilliance of molten rock as it breaks through the earth's crust in a volcanic explosion. *Liberty* in the gentle pinks, soft grays, and dusty yellows of the egg tooth as it pierces through the chick's shell in an oxygen-delivering pip. These sorts of cracks are definitive; should an infinitely long period of time elapse, the surface may still not have reformed. Seed coats, egg shells, and earth mantle do not primly mend with age-defying stitching, conscientiously brushing themselves off and lifting up their hopeful faces to shine anew. No. Permanent freedom dresses in an altogether different fashion: a thread that ladles reservoirs of adenosine triphosphate, releases megatons of thermal energy, and gyrates an embryonic body for days.

Cracks conduct movement and advance entropy. The crack of curiosity allows the hidden to fly with hatchling wings, the unthought to effuse in a lavalike flow, and the buried to root through nutrient-rich soil. A rooting radicle. A radical conception. A thought uprooted, rerooted, changed from the root as radicalized. A riddled similarity, the curious mind and the first fruition of the seedling are notably alike. As naturalist Charles Darwin penned in his *The Power of Movement in Plants*, "It is hardly an exaggeration to say that the tip of the radicle thus endowed . . . acts like the brain . . . ; the brain being situated within the anterior end of the body, receiving impressions from the sense-organs, and directing the several movements."[2] Tucking their head down in the safe darkness and anonymity deep beneath the turf, the radicle riotously flings their white sinewy arms into the superficial dirt and spreads their green leafy legs into the air. And then they walk. Walk the air. Walk the sky. Walk over the face of the sun. They walk a circumnutating trail and think—think about how the world looks from this angle, from that distance, from above, beneath, and aside. Their imbibition is a conduit for imbrication; the crack in the soaked seed coat from which the radicle first peers is the prerequisite for the curious walk by which edges of distinction overlap, layers of perception form, and networks of realization grow. When next it rains, we must pause beneath the sky to soak our woolen coats. Then we will stand on our heads and step our legs across the clouds.

*   *   *

We read on, our identical right pointer fingers feeling the smoothness of the paper as it gently moves down the column of "C" entries in Curiosity's dictionary:

> (*vb.*) to be cracked (induced); to be opened, as in a front door, a molecular pore, a privy closet, or a memory box.

Any hinge is a crack in potentiality, an opportunity to separate two abut-
ting bits from one another, a chance to find the line of demarcation and
pry apart. Unlike the permanence of rupture, fracture, and explosion, these
cracks are meant to be induced and then reinduced; we crack a window to
feel the light spring breeze, and then we crack it again to smell the crisp
note of snow, the soft air of rain, or the silly frolic of pollen. A seasonal
hinge. As the rotating crank handle drives open and then closed the panes,
so the spinning of the heavenly bodies drives open and then closed a celestial hinge, permitting the sun's rays to ricochet off the moon, brightening
our night. At the full moon, the crack is open; at the new moon, the crack
is closed. As the moon sets, the sleepy leaf pivots on its cellular hinge,
the pulvinus, turning from its quiet communion with others to a vibrant
exploration of the air, stretching out and away. Like a child, it reminds us
of its independence each morning. In the daylight, the crack is open; in
the moonlight, the crack is closed. Through days, months, and seasons,
we have the opportunity to curiously crack open or creak shut the hinged
parts of life.

By cracking open, we perceive the scent of the breeze, the light of the
sun, and the wide vista of an open leaf. By peering through the open angle,
we see the forgotten, the useful, the treasured, or those that might become
so in the future. These are the materials about which we fix a hinge or place
a lid. We grasp and we gather, from the annals of nature in, around, and
among us. In the *Kalevala*, an epic poem compiled by philologist Elias Lönnrot in the nineteenth century from Karelian and Finnish oral tradition,
we learn runes from the cold and songs from the wind. We take the "sentences the trees created," roll them together into bundles, and fasten them
onto our sledges. When we have finished gathering, we head for home.
There we unpack, tying them to the rafters, hanging them from the door
frames, draping them over the exposed beams. We wonder whether some
sentences reach the floor, or traipse along the rug and out the front door.
We wonder whether sentences are transparent like sunbeams or translucent
like dust streams, and whether they look or feel like willow branches cloak-
ing our bare shoulders. After they've hung, we lay "them in the chest of

boxes / Boxes lined with shining copper."[3] There they stay, safely protected from the elements, hidden from the eyes, treasured in the mind. Until one moment, one day, one month, one season, when we have forgotten ourselves, or perhaps when we are hopeful to learn anew, we crack open the carved lid of the old trunk. We smell the scent of the trees: the grains of the chest without, the woodiness of the words within. The light of the room falls into the copper-lined darkness. The sentences stretch and open like leaves.

*   *   *

We close the dictionary, watching the sentences curl up. We open it again, watching the sentences unwind and stretch. We close and open, close and open, again . . . and again, noting the peculiarity of the sentences' movements. And then, eventually, we read on:

> (*vb.*) to be cracking (voluntarily); 1. to be sounding an opening, as the calving face of a glacier, or (more humbly) a bowl of Rice Krispies, or (more honestly) paper being crumpled; each voice calls out its own cracking. 2. to be doing the opening, purposefully, as a jaw at the joint, a finger at the knuckle, a hand at the palm, or a leg at the knee.

While some cracking is permanent or induced, this type of cracking is repeated and volitional. Across the past, the present, and the future, cracking is a way of life, a manner of being. By choosing to crack curiously, we open the mind, move the body, and prime the senses. For the Venus flytrap and the humpback whale, the choice is to crack open the mouth, wait for prey, clap shut, and enjoy. For insects and scallops, the choice is to press the wings or valves at the hinge to drive the animal through the air or water, whether in search of food or safety. For the rabbit's ear and the human's eye, the choice is to bend toward and be receptive to a sound or light, a movement or being, an event or curio. Ears and eyes, yes. But really those eyes! The hundreds of blue eyes lining a scallop's mantle. Each sits atop its own extendable tentacle and contains crystal mirrors composed of millions of square tiles; each telescoping to perfectly reflect every photon of blue light. Eyes are cracks that mirror and they are cracks that vent. The eye is a crack forward into the world, but also backward into the soul; it is a gate of inquiry, an invertible vision: the horizon where the space of them and the earth of us meet in an unruly tangle along the line of the lid. What do we see when we crack our eyes inward? We peak in, peep on, and snoop about. While eyeing, we blush for nosying. It's a raucous place. Words live here, "variously and strangely, much as human beings live, by ranging hither and thither, by falling in love, and mating together," writes Virginia Woolf.

> They seem to like people to think and to feel before they use them, but to think and to feel not about them, but about something different. . . . They hate being useful; they hate making money; they hate being lectured about in public. In short, they hate anything that stamps them with one meaning or confines them to one attitude, for it is their nature to change. . . . When words are pinned down they fold their wings and die.

When we look within, we see words that maintain an authenticity to themselves and their curious need to change; they ask us to mirror, not to conceal. To mirror and to vent. While living in the mind, words hang together in questions, "sensitive, impulsive, and often foolish," that "clamour" to be asked, to be vented "openly in public." And yet they are particular about what kind of public, knowing that "power and prestige come down upon" precocious questions "with all their weight." Questions "shrivel up in an atmosphere of power, prosperity, and time-worn stone. They slink away to less favoured, less flourishing quarters where people are poor and therefore have nothing to give, where they have no power and therefore have nothing to lose."[4] When we look within, we see questions that demand an audience both with us and with them, in a space where power structures are dismantled and constraints of prestige are subverted. Yes, here is a boisterous place of possibilities that we need only crack our eyes to see. To see, to reflect, and to pass on.

*   *   *

We look up from the page. "Are you two reading the dictionary again?" we are asked.

"Naturally," we reply in unison. We decide to flip to the back of Curiosity's dictionary, where the verbs finally end and the nouns begin. We look at one another. "Here's to cracks, then!" our eyes say.

(*n.*) the crack in ourselves. (Entry yet to be written.)

Some of us are much like the average tortoise: we pull our shells up to our neck each morning and practice a few head scrunches in case further retreat might prove necessary later that day. But all of us, really, live inside ourshells. Shells of habit and self-perception. Of social expectation and failures of imagination. And these cages of calcium carbonate are by turns tenderer and tougher than their compatriots. Cracks come commonly for some, and once in a blue-tinted moon for others. Cracks in our faith and in our facts, cracks in our body's capacities and itineraries, cracks in our companions and what glues us to them. Cracks in the very geographies of ourselves. But that is precisely what is needed.

Gloria Anzaldúa invites us to find our wells and our whirlwinds, our *cenotes* and *remolinos*. Where what lives has yet to find home in a distinguished set of words sitting around a distinguished table of vittles nourishing a distinguished lineage of knowers. When we crack, creativity can happen. When we crack, impropriety can happen. Without that vulnerability, that border-thinking, *those questions*, we will be stuck on our own islands (or perhaps what's worse, stuck up on the mainland), without the capacity to bridge, to paddle about, or even to leap into the sea that lends everything an irreverent (and nonrevelatory) context. Let those "liberating insights/*conocimientos*," Anzaldúa writes, "burst through the cracks of our unconscious and flow up from our *cenotes*." Let them crack us open to new identifications, disrupting our categories, showing the flaws in our social cultures and the faults in our preconceptions. Questions disorient and dislocate. And in so doing, they make new selves and nonshells possible. "There is a crack," she reminds us, "a crack in everything / That's how the light gets in."[5]

*   *   *

We turn the page, gingerly because dictionary pages are so thin. Why is it that thin pages make more sound than thick? A bit of sunlight glints off the wind chime and draws a bright dash across the next entry:

(*n.*) the crack in our world. (Entry still blank.)

It is not merely ourselves that encircle us but also our worlds. There are grand architectures within which we make our moves and vast webs across which we slink. There are known histories and there are submerged histories, forming icebergs all around us. Rivers of beliefs, mountains of books, and deserts of policy briefs—not to mention the old, old forests of blunders and biases—surround us the whole world over and for eons too. Unseeable things and unsayable things that nevertheless define in advance how it is we are born, how we grow, what we do, and where we die. How do we find the crack in our world and walk through it? How do we break and break open everything that is hollow? Break and break open everything that has been so long hallowed? By what measure does "might" become plausible?

For Friedrich Nietzsche, we must set ourselves round and stretch ourselves long, and draw down the lightning crack of a question mark over easy or simplified answers. We must invite the sudden flash of sky-fire that illuminates the fault lines in our familiar, sometimes forced systems of facts. And the bald wind that brushes free the seams of what is seeable and sayable. And the pelting rain and rivulets of erosion, against which no certainty can stand. *Fragezeichen*: question mark. Nietzsche would call it the question mark that marks all things as questionable.[6] The crack that opens our world. There is no telling when these "strange, wicked, questionable questions" will strike.[7] But would that they struck more often, and ever closer to those fundaments we have yet to realize stand unceremoniously and unjustifiably overhead. What better way to throw off the weight of such things than to weigh them down with a question mark, slung like a leaden anchor round the neck of an idea and just watch it sink? Such might prove to be "a schooling in suspicion," but just as equally an education in "courage and audacity."[8]

*　　*　　*

We marvel at the large, empty white space on these pages. Aren't most dictionaries full to the brim? Spilling over, text size shrinking, magnifying glass needed? Why print a dictionary with terms listed and no definition provided? Questionable practice that.

    (*n.*) the crack on the horizon. (Blank.)

Cracks can be courted and romanced, squandered and lost. But one crack never leaves so long as we breathe, and that is the long, shallow crack on the horizon. The horizon is, after all, a crack between the worlds of things underfoot and things overhead, where the sky and the earth meet, but do not intermingle. Where the Between beds down permanently. The horizon beckons us onward, betrays our settled assumptions, suggests there is always more than we know and more ways in which what we know can be thrown into curious disarray. Having traveled the world over doesn't make this not so. For there is a structure of mystery to the horizon that is irrevocable and therefore faithful. And it is upon that horizon that any hope of the future, as precisely *not* the present, hangs.

Phenomenologist Maurice Merleau-Ponty speaks of the horizon as what keeps our questions open. The horizon is an edge that opens onto something else, something more. It is always there and yet it can never be reached. In a sense, everything has a horizon. As such, no inquiry— and certainly no object of inquiry—can ever really be closed down, finally decided, boxed up and indexed. There is always a bend in the road, a corner to turn, a perspective yet to be taken. This keeps our quests—as much as ourselves—open to transformation. And, oddly enough, that unites us. "No more than the sky and the earth is the horizon a collection of things held together," Merleau-Ponty writes, "it is a new kind of being, a being by porosity . . . and the one before whom the horizon opens is caught up and englobed within it."[9] The horizon is not simply an edge that drops off but also an edge that connects. Its very uncertainty draws us in. Insofar as the horizon is always on the edge of something, it is also the precondition of all edgework, all sense of belonging and context. May our questions, then, be ever born in this nexus of belonging and becoming. Let us be curious in the space of our shared questionableness, always present to what is and ready for what is next.

*　　*　　*

On the few completely blank pages at the end of Curiosity's dictionary, we pause. We take out our identical black pens and slowly begin to write.

(*vb.*) having (*n.*) crackability; portending the possible.

Plant seeds have crackability, as do hinges and eyes, sentences and queries. Many of us are often afraid of our own crackability—and indeed the crackability of our worlds and our horizons. We fear, perhaps rightly, that it signals a deep capacity for loss. We would rather avoid being unsettled. We worry about being remade. The disorientation inherent in crackability is deeply disconcerting for creatures who spend their days crafting settled things. But crackability can be modeled, and crackability can be learned. Look again to the small seedling, and try channeling its willingness to break open and to give way to growth and beauty and new forms of life. In the cycle of what was and what will be, crackability is key. And this is among the most innocent of briefs for curiosity.

To court crackability, one must be willing to be severe but also silly. Consider the sacoglossan sea slug. Known for their sap-sucking and sun-guzzling habits, as they feed from nearby algae, the sacoglossans are also famed for their crackability. When for whatever reason a sacoglossan acquires a parasite or some other disease, the slug cracks! Its neck becomes a "breakage plane" and its body falls clean off. Severe indeed. But also silly. Slugging about with barely its head and a bit of antennae, the sacoglossan strikes a comedic pose. But there is seriousness here too as it regenerates its heart, limbs, and everything else. What no longer serves it? What detracts from its flourishing? And what jeopardizes the health of sea slugs everywhere? The sea slug manifests a certain willingness to loose (and lose) its node over and over again and to regrow its edge. It is a curious willfulness to throw off familiar ways and look at the same thing in a different way.

As such, Socrates might well find in the sacoglossan an unexpected companion. Imagine the two out about town in Athens, extolling the virtues of crackability.

In the human world, we can learn a lot about crackability from people who have had to crack open spaces for themselves and their own queries. What is feminism, for example, if not the cracking of patriarchy, its myths and foundations?[10] Musing upon a feminist ethos, scholar and writer Sara Ahmed retells the story of a girl whose curious desires and doubts were interpreted as rebellious willfulness.[11] We can imagine her hand constantly shooting up above her head. Grimly, her recalcitrance led to her deathbed, but, upon her burial, her hand shot up through the ground regardless. It is a story of the crackability of the earth, the hinge between life and death, and indeed the irrepressible seed of willful curiosity (and curious willfulness). Children often see the cracks in our facades and the seams (seems?) in our social mores better than we do, having been incompletely acculturated. Likewise, too, do people marginalized within a given social system—those for whom it was not made and for whom it does not work. If we are to court crackability, we need everybody. And we need a new architecture of listening by which sidelined voices slide forward.

There are innumerable totalizing structures that leave little room for cracks and for curiosity—indeed, that deny crackability outright. Long traditions of what counts as a knower, what counts as a system of knowledge, and what counts as a way of knowing. But curiosity invites us to lose count and to count again in the crack between nodes and nodal hubs. In his book *In the Break*, poet and scholar Fred Moten reflects upon the Black radical tradition's capacity to resist and to reimagine. He thinks about jazz improvisation. And he thinks specifically about the break: that moment when the music is suspended, and a percussive or instrumental solo cuts in. The piece is broken open, and its motifs and rhythms are taken in unexpected directions. "This refusal of closure," he observes, is "an ongoing and reconstructive improvisation of ensemble," of thinking with and through one another. Appreciating the creative power of this caesura, Moten writes, the Black radical tradition mobilizes what "slips through the cracks" and practices "an ongoing shiftiness."[12] This is resistant invention. This is fugitive curiosity.

Curious futures have crackability. The undoing that allows for greater doing, the disconnections that allow for new connections.

*   *   *

We close the book. "Here's to the cracks in our curious futures!" we say—wordlessly, in our characteristic nod of the head. We open the front door (or is it a back door, side door, or skylight?) and head out.

Appendix: A Curious Bestiary

In the spirit of multiplying models and practices of curiosity, we offer here a brief bestiary. We catalog a few of the animals that have been thought especially curious (or especially good models of curiosity) over centuries of especially Western thought and throw in a few of our own. We recognize that any such compendium will always be incomplete, and in multiple ways. Nevertheless, we offer it as an opening. In attuning ourselves to these animal lives, we tend to their distinct knowledges and ways of knowing, we celebrate their diverse curiosities, and from them we draw inspiration to honor the animal curiosity within us and beyond us.

Ass: Have you ever wondered what it would be like to be a duck? Or perhaps a dragonfly or a chipmunk? Such wonderment is a time-honored tradition. Lucius, the protagonist of the first fully preserved Latin novel, Apuleius's *Metamorphoses*, always wondered what it would be like to be a bird.[1] And not just any bird—not a pigeon or a canary, a robin or a titmouse, no, but an eagle! Ever so "curious," Lucius sought out women who were skilled in the magical arts and might oversee the transformation. Unfortunately, however, the spell went wrong, and he was turned into an ass. Neither sleek nor aquiline, Lucius became a lumbering beast with thick hoofs and a broad snout. Used to circulating among the wellborn and the well-bred, Lucius was suddenly thrown into poverty, illegality, and ill-starred fate. Lucius's tale, then, begs us to ask a series of questions. What might poverty teach curiosity? How might curiosity be transformed by humility? What do lack and loss make questionable (and unquestionable)? What shape does curiosity take in the mundane bluntness of the everyday? And indeed, what does struggle train in the curious heart, whether one finds oneself beating one's wings or pawing tenaciously at the earth?

Bee: Characterizations of the human species are not always kind. Nevertheless, in a moment of largess, Friedrich Nietzsche offers one. We are "by nature winged creatures and honey-gatherers of the spirit," he writes.[2] We collect knowledge, we treasure it, and we build our beehives full of it. "The brain is like a comb exactly laid," seventeenth-century philosopher and scientist Margaret Cavendish remarks, such that every thought lies "each by itself within a parted cell."[3] With an average of a hundred thousand cells per hive, we are kept quite occupied. But in our buzzing and our busyness, learning this and discovering that, Nietzsche warns, we often fail to settle into the present. We rarely get around to asking basic questions of ourselves like, "Who are we really?" Of course, the wonderful thing about bees and their pollen-collecting trails is that, in foraging food for their colonies, bees make plant life itself possible. They work communally among the gatherers *and* among the gathered. From this vantage point, curiosity seems not to warrant Nietzsche's worry. Far from a consumer-capitalist track that contributes to widespread self-alienation, the bee's curiosity is relational and respectful, giving back as many nourishing opportunities as it takes. Winged creatures, honey-gatherers, givers.

Bookworm: "They smell so tasty!" exclaim our Noses; "So delicious!" rave our Tastebuds. And beetles, termites, ants, moths, cockroaches, silverfish, book lice, and book scorpions agree. Books taste delicious in the mouth and mind, feel delicious in the hand and claw, and look delicious whether lined up neatly on a shelf or left open on the floor—haphazardly strewn, mid-chew. We want more, and so we find more. We grasp them in our palms and hold them to our hearts. At times we even collect them, . . . or we might wish to. "Why, in five centuries, in six countries, do there seem to have been so few women book collectors?" asked Mary Hyde Eccles at the Grolier Club's 1990 exhibition *Fifteen Women Book Collectors*. "The answer is obvious: a serious collector on any scale must have three advantages: considerable resources, education, and freedom."[4] We celebrate the few: Diane de Poitiers; Catherine de' Médicis; Gabrielle d'Estrées; Christina, queen of Sweden; Madame de Pompadour; and Amy Lowell. But might we all be collectors in some more common sense of the term? Collectors-cum-ingestors, gathering up the tastiest bits. "She has a robust appetite," pens Virginia Woolf, describing seventeenth-century writer Madame de Sévigné, "a natural dwelling-place in books . . . not read by her so much as embedded in her mind."[5] Books become a bit of our inmost selves, just as an apple becomes a bit of a child's eye. We are nourished by the pulp of woody words, filled by the glue of spiny sentences, satisfied by the binding of leathery language. The text becomes the tissue of us, but more than that, it lives on within us. "Both bodies and texts are constantly becoming," writes Emily Rose, scholar of literary translation.[6] And the body nourished by text has a hybrid becoming. As a food never fully used, strands of meaning from a long-forgotten page may lie dormant, unseen, unknown for years, always with the potential of visiting us from within. Someday—perhaps today—we may notice them in the glint of a hair, the pulse of an artery, the crook of a knuckle, the quickening breath of a beckoning experience.

Canary: Although a common account of curiosity is as self-ignited, a common *experience* of curiosity is as other-ignited. An internal fire lit by another's stray spark, inflaming a new passion. To US author Ruth Stiles Gannett, this infectious, contagious, transmissible curiosity is canary curiosity. Flute is a canary combusting from Feather Island's disease. The ruler King Can the XI is "actually dying of it." It all started with King Can I, who was never well or happy. When his subjects asked him why, he would simply say, "I'm dying of curiosity." The canaries were curious to know why he was so curious and so they quickly also became sick with curiosity, and each successive generation eventually died of it. Elmer, a boy traveling with a blue-and-yellow-striped baby dragon with burnished gold-colored wings, wondered "what they could have been so curious about." To which Flute retorted, "See, there you go getting curious! What a great day it will be when this island gets over the plague of curiosity!"[7] But rather than any doom materializing, the infected Elmer joins the canaries in excavating new lines of knowing that celebrate bravery, generosity, and belonging. What new hopes might burn from sharing our current contagions?

Cat: If you know anything about curiosity, you know it killed the cat. And that means, you know this: curiosity is risky. It is dangerous. Even nine lives are not enough to survive its hazards. The saying has a remarkably long and storied history, traceable all the way back to playwrights of the sixteenth century (and the luminous William Shakespeare). Curiosity will catch you unawares and shatter your expectations. And that can be fatal—especially for cats. But what if curiosity were not the culprit nor death the sanction after all? What if, as poet Alastair Reed puts it, the cat was simply "curious to see what death was like"? After all, "dying is what the living do" and "dying is what, to live, each has to do."[8] In this sense, dying is not some regrettable repercussion or karmic comeuppance but a natural object of exploration. Day in and day out, cats are on the alert, their whiskers forever detecting the slightest vibrations of air and matter. Dying is but another form of attention—less risky than it is merely exploratory. A way of vibrating differently.

Dipper: Nothing feels quite as refreshing as a cool dip on a sweltering summer's day. When we feel hot and stuffy in our heads, we often dip into a blog, a podcast, a book, a track, a reel, or a conversation. We splash about, take a plunge, drift and float. Like the water ouzel (or dipper), we are stocky, dark-gray birds that dive under water, walk, and fly beneath the surface, coming out with a dragonfly nymph, crayfish, caddis fly larva, or tadpole for lunch. In her book *The Dipper of Copper Creek*, writer and naturalist Jean Craighead George follows a dipper as he "skimmed the grinding ice floes and flew through the bursting sprays as if the stream had created him, and were reluctant to release him to the sky. Then he went back where he seemed to have sprung from, the lashing white waters of the opening creek."[9] This water is where we have come from, where we are going, and from which we will burst forth like sea spray. After swimming and swimming and swimming for what feels like hours, we come out "thoroughly exhausted" and "completely soaked," like Milo and Tock in *The Phantom Tollbooth*, incredulous that the Humbug (and others like him) "can swim all day in the Sea of Knowledge and still come out completely dry."[10] We find the wetness exhilarating. That feeling of the water becoming part of us, the knowledge changing us. After a while, our mind's fingers look like wrinkled prunes, with which we can more easily grasp underwater objects: those slippery thoughts covered in bothersome barnacles, those ruminative rocks coated in diatoms and blue-green algae. Before we grasp them, we first *see* them thanks to the dipper's transparent third eyelid, opening up the underwater panorama. And when we surface, we sing a song "heard by none," for we sing "to the accompaniment of the pounding waters of the mountains."[11] How might we next dip, get wet, see through a different lid, and sing for no one?

Fly: What distracted dithering and banal buzzing! Saint Augustine worried that just outside of proper reflection, there is a kind of curiosity drawn to distractions such as "a lizard catching flies or a spider entangling them in her nets."[12] Such a vain interest in light and insignificant things refuses to root itself in divine value or pragmatic use. It is curiosity for curiosity's sake. René Descartes, however, took great pleasure in precisely this practice. Preferring to sleep late and lounge in bed for hours, he would watch flies scuttle-hop across the ceiling. As legend has it, he developed the Cartesian coordinates (or the x and y axes) as a way of mapping where the flies would land. Socrates, too, flouted propriety when he became a gadfly to the Athenians, lobbing insistent, irritating questions at them to unsettle surety and illuminate ignorance.[13] Like a winged insect goading its host, the ignominious Socrates woke Athens from its sleep. Could this be, in each instance, a case of dipterous curiosity? And what happens when dipterous curiosity is let loose in the university, Woolf wonders. The sort of "three-legged fly" that has "survived the winter" of a dreary, bone-cold scholasticism whose prestige mongers refuse to let in the light?[14]

Frog: Images are powerful communicators. Working in the early seventeenth century, Italian iconographer Cesare Ripa created perhaps the most memorable image of curiosity before or since.[15] He drew a woman standing boldly against a bare landscape. The hair on her head, the wings at her shoulders, and the arms at her side all lifted to the sky. She is striking and elegant, but for her robe, which is, quite unexpectedly, covered in large ears and still-larger frogs. Not images of ears and frogs, mind you, but the things themselves, hanging there heavily as if ready to be plucked or to leap free of their own accord. This, Ripa insists, is the spitting image of curiosity. Her eagerness to hear and to see the newest news is undeniable. She's lost in the moment. But frogs? Why frogs? Frogs, he explains, are special "emblems of inquisitiveness . . . by reason of their goggle-eyes." Frogs do in fact have unusually large eyes, and for good reason. They enable them to see out in front, to the side, and a bit behind. What if a frog-like curiosity, then, was not newsmongering so much as contextualizing, situating us spatially, but also temporally? A certain simultaneous eyeing of the past, the present, and the future?

Grasshopper: You know that feeling when you simply must be on the move? In your mind, your ears, the tips of your fingers, you hop, jump, leap, and spring. Sometimes—as US author Arnold Stark Lobel well knows—your latest curiosity road trip is interrupted by three seemingly identical bright-pink butterflies with wide, vacant yellow eyes and perfectly curled proboscides. They forcibly invite you to become a part of their daily life:

> Each and every day we do the same thing at the same time. . . . We wake up in the morning. . . . We scratch our heads three times. . . . Always. . . . Then we open and close our wings four times. We fly in a circle six times. . . . Always. . . . We go to the same tree and eat the same lunch every day. . . . Always. . . . After lunch we sit on the same sunflower. We take the same nap. We have the same dream. . . . Always. . . . When we wake up, we scratch our heads three more times. We fly in a circle six more times. Then we come here. . . . We sit on this mushroom. . . . Always.

They tell you that they will meet you at that exact mushroom every day, and that they will tell you about their scratching, flying, napping, and dreaming, and that you will listen, "just the way you are listening now." After you politely decline, citing that you will be "moving on" and "doing new things," they ask incredulously, "Grasshopper, do you really do something different every day of your life?" "Always," you reply, "Always and always!"[16]

Hedgehog and fox: An ancient Greek poet from the island of Paros, Archilochus famously penned, "The fox knows many things, but the hedgehog knows one big thing."[17] Desiderius Erasmus later Latinized the line: *Multa novit vulpes, verum echinus unum magnum.* British children's author Beatrix Potter brings these ancient icons to life in depicting Mr. Tod the fox as having "half a dozen houses," but being "seldom at home," whereas Mrs. Tiggy-Winkle the hedgehog is the sole provider of clean clothes to the entire animal community, while her nose goes "sniffle, sniffle, snuffle" and her eyes go "twinkle twinkle."[18] For centuries, people have debated which is better, the fox or the hedgehog, and who is who. Some say essayist Michel Montaigne is a fox and Plato is a hedgehog, while others say the sciences are the fox and the humanities are the hedgehog. Still others suggest we ought to be both: the mystical foxhog, curious to know one and many simultaneously. But perhaps in a moment of quiet reflection, we might realize that in knowing many things, we are forced to know really only one thing, and in knowing one thing, we come to know many things. Just as the fox and hedgehog belong to a larger ecology, so too does knowledge have its own cycle of life. The more we focus our curiosity, the more diffuse our ignorance becomes, just as the more our curiosity wanders, the more likely we are to find a through-line.

Hound: An eminently respectable, even regal creature, the hound is trained to track and to chase. In order to hunt, the hound has to hold a single scent in a saturated scene, a bit of olfactory fidelity in a riotously sensuous world. In his ruminations on curiosity, Roman essayist Plutarch uses the hound to personify a focused inquisitiveness. Hounds, he remarks, do not "turn aside and follow every scent" but instead "keep their sense of smell pure and untainted for their proper task."[19] They save themselves, so to speak. And while this ascesis clearly equips them to locate—and in some cases to kill—it can also do something else. It allows the hound to listen and to attend and to submit their existence to the life force of another. It humbles the mind and clears the cobwebs so that the act of following becomes possible. What curious practice is this that gives someone the chance to find only insofar as they follow? Might the hound teach us that curiosity requires a certain renunciation of control and a certain commitment to companionship?

Inchworm: Curiosity need not always be dramatic and expansive. Sometimes it is unobtrusive and very, very small. Like an inchworm, we might question locally, move nearby, and progress slowly. We might ask something others think is too small to ask. But with our littleness and our minute movements come special powers, as is well appreciated by the Miwok, a group of four linguistically related Indigenous tribes who live in what is now Northern California. In various tellings and retellings, the Miwok recount the legend of Tu-Tok-A-Nu'-La, which means "measuring worm." Laying down to rest in the sun after swimming in a creek, two boys fell into a magical sleep. As they slept, the ground on which they lay rose higher and higher into the air. The granite continued to rise for years until the boys' faces "brushed the sky" and "scraped the moon."[20] The animals took council with one another, and determined to save the boys and return them to their mother. Each animal attempted to scale the granite wall. Where mouse, rat, raccoon, grizzly bear, and mountain lion failed, the inchworm succeeded "step by step, inch by inch," recounts Julie Parker, born to a family of the Coast Miwok and Kashaya Pomo tribes. And as the inchworm steps, the inchworm sings, "Tu-Tok, Tu-Tok, Tu-Tok-A-Nu-La; Tu-Tok, Tu-Tok, Tu-Tok-A-Nu-La."[21] Reaching the top, the inchworm invited the boys to sit on his back, and then he "led them on a continuous, circuitous slide" to the base of the wall that now carries his name (also known by the Spanish name El Capitan) as they "laughed and screamed with delight."[22]

By inching, the measuring worm reaches spaces that others cannot and builds community where others fail. Local questions, small spaces, and the fine textures of rock face that go unnoticed by others, invisible to the broader society, invite us to step, to walk, to inch curiously and relationally.

Monkey: Perhaps paradigmatic of inquisitive creatures is Curious George, a beloved children's book character who, since the 1940s, wanders about the world exploring, experimenting, and ultimately getting into endless trouble.[23] He is unable to live seamlessly in human society and generally throws a wrench into every bit of its machinery. Perhaps most curiously, however, George is never curious about his colonial past—that moment when the white man in the yellow hat scooped him up from Africa. As such, George's irreverent, more-than-human curiosity is a curiosity at the edge, but it is nevertheless unable to catch sight of the sociopolitical context within which it swims. Intriguingly, flesh-and-blood monkeys are relentlessly curious about counterfactuals—that is, how different choices produce different outcomes.[24] Might it be, then, that monkeys themselves invite Curious George to be still more curious, to wonder where he came from and what could have been otherwise? To get up to some counterfactual trouble? The sort of ruckus only possible with a decolonial imagination?

Octopus: Much distinguishes the octopus. Aside from its three hearts, the octopus has nine brains: one between its eyes, and one in each of its eight tentacles, themselves lined with as many as 2,240 suction cups. We humans typically (although not always) have but one heart, one brain, two arms, and ten finger pads. It is no surprise, then, that the octopus seems the perfect figure of a relentlessly open-hearted, open-minded, open-armed curiosity, bent on collecting new information from every corner of coast and crevice. Paola Antonelli, a senior curator in the Department of Architecture and Design at the Museum of Modern Art in New York, describes herself as "a curious octopus," explaining, "I am always reaching out and taking in things from everywhere."[25] No doubt an important curatorial instinct. But the octopus uses its warm, intelligent "hands" not only to catch and to collect, but also to swim, to cut a caper, to dance in the reeds, and even to play among nearby schools of fish. What curious activity is this, these swinging and swooshing arms that compel and contextualize and frolic?

Rabbit: Alice—for she was not yet Alice *in* Wonderland—sat bored on a riverbank, with little to do and still less to think. That is, until she saw a white rabbit in a waistcoat nervously mumbling and checking his time-piece. "Burning with curiosity," Alice dove after him and all the world turned upside down.[26] Almost 150 years later, Neo, in a similar state of lassitude, sees the white rabbit on Dujour's shoulder and falls headfirst into the Matrix. Directors Lana and Lilly Wachowski confess the story is a thinly veiled commentary on gender transition; the Matrix is a "trans metaphor."[27] Rabbits, then, are a short hop from Red Queens and red pills. But why rabbits, in each case? Rabbits prefer not to walk. Not the sort for an anxious amble or furtive footstep, rabbits prefer to hop. Darting out of danger or bounding into the underbrush, rabbits move by a series of leaps, launching from their hind legs at once up and over. When happiest, they add a little twist to the height of the hop. As Alice and Neo came to know, certain kinds of curious change, certain travels between dimensions of being, require an unusual practice of propulsion and pleasure. Making a habit out of the hop.

Robin: With its dark, questioning eyes, the robin sits in the mind of every child, staring, lost in thought, peering out a window. British American novelist and playwright Frances Hodgson Burnett notes those curious eyes in her book *The Secret Garden*. When Mary asks the gardener of Misselthwaite Manor, Ben Weatherstaff, what kind of bird was following them, Ben replied, "Doesn't tha' know? He's a robin redbreast an' they're th' friendliest, curiousest birds alive. . . . He's always comin' to see what I'm plantin'."[28] The robin's curiosity looks to see, and sees to know. He wants to know what plant is joining his world, what new leaf and what new flower will decorate the garden of his home. By hopping along the meandering walkways, the robin joins the other inhabitants of the garden in curiously growing his knowledge "*along* the paths they tread," as anthropologist Tim Ingold puts it.[29] The narrator in *The Secret Garden* continues by observing that "the robin hopped about busily pecking the soil and now and then stopped and looked at them a little. Mary thought his black dewdrop eyes gazed at her with great curiosity. It really seemed as if he were finding out all about her."[30] Might robin-like curiosity acknowledge the times when what must be known can only be seen? There is no question to ask. No sound to make. No boisterous action to perform. One must simply be still and look on, quietly.

Snake: The snake slips in and out of the understory. Sometimes it slinks along the ground, sidling between grasses, over river rocks, or under shrubs. Other times, it slithers up tree trunks, winds around their branches, and hangs for a bit in lazy loops. In the Judeo-Christian tradition, the snake—or the serpent, to put it more sinisterly—is the icon of curiosity.[31] It was the serpent, after all, who tempted Eve with curiosity. He whispered in her ear, winding through the byways of her mind, slipping here and hanging heavily there, making shapes in the space between things. And with that space, Eve reached out her hand to grasp the forbidden fruit from the tree of knowledge and all the world fell into darkness. Accused of being sly, deceptive, even seductive, however, the snake is perhaps better thought as simply slippery and, as such, as having the enviable ability to curiously plumb the slipperiness of things around them, whether things thought, things said, or things found in and around the ground. Gloria Anzaldúa said as much when she characterized the snake as a figure for "more life," the sort of writhing, wriggling movement that deepens an awareness of a restless world and ultimately "fuels [its] transformation."[32]

And how better to plumb a slippery world than to have a flexible, filamentous body, searching as it lives—elusive, difficult to hold, and impossible to stand on? We've seen this relation between form and function before in the butterfly antennae: a long, thin bodily shape and a searching, curious way of being. Maria Sibylla Merian, seventeenth-century botanist, noticed the striking visual similarity between the snake and butterfly antennae. Regarding one species, she notes that "its antennae look like two colourful snakes, white and black," whereas for another she writes that "the antennae are like black snakes."[33] Might snakelike curiosity further embody slipperiness in the thin antennae-like lines of inquiry that strike out from oneself in filamentous questions or filamentary questioning? Tendrils of fauna; vines of animalia. Snaking out, snaking round, snaking through.

Acknowledgments

A romp of this sort prompts all sorts of reminiscing. We cannot help but recollect those with and against whom we have found the sweet freedoms of curiosity. As we extend our thanks herein to certain folks, we also want to mention those unnamed: the writers and scientists, both of flesh and of spirit, without whom our path would have been impoverished or even impossible. Curiosity for us has never been a solo production, despite our propensities for the solitary. It has been a project of friendship and communion. May this be but the beginning of our testimony.

We wish to thank the students in our curiosity classes, including the Curiosity: Ancient and Modern Thinking about Thinking course at the University of Pennsylvania (fall 2018 and spring 2020), the Ethics of Curiosity course at Hampshire College (spring 2016), and the Philosophy of Curiosity as well as Curiosity, Politics, and the Public Realm courses at American University (spring 2018 and fall 2019). Without students' incisive questions and rich lines of flight both within and beyond this material, this book would not be half as interesting.

We are grateful to the many audiences with which we have discussed our curiosity-related research, stretching around the globe. Specific thanks goes out to the American Educational Research Association, American Philosophical Association, Critical Genealogies Workshop, Derrida Today, Diverse Lineages of Existentialism, philoSOPHIA, Society for Phenomenology and Existential Philosophy, and Trans Philosophy Project conferences as well as audiences at Academic Programs International, Adelphi

University, Affective Brain Lab in London, American University, Appalachian State University, Brooklyn Public Philosophers, Centro de Ciencias de la Complejidad, *Choose to Be Curious* podcast, Complex-Space—CSS Satellite, DePaul University, DePauw University, Five College Women's Studies Research Center, Hampshire College, Haverford College, MIT Media Lab, University of Colorado at Denver, University of North Carolina at Charlotte, Imagination Institute, Penn Network Visualization Program, Lea Elementary, Open Connections, Technology in Psychiatry Summit at the McLean Institute, Trinity College Dublin, Villanova University, Westtown School, and Wikimedia Foundation.

We offer appreciation to our teachers and professors for igniting our curiosity, especially Sandy Feinstein, Thomas J. Lynn, David Mills, Michael Naas, and Emily Zakin. A special thanks goes out to Barry and Mary Hannigan for introducing us to J. R. R. Tolkien along with the wonders of Johannes Brahms and Frédéric François Chopin.

We wish to thank the minds of those people long past who wrote books and poems that continue to nourish our thinking, whether housed in the Borders bookstores of our youth, or more recently, Baldwin's Book Barn, Myopic Books, Raven Used Books, and the Montague Book Mill.

We extend our thanks to all our close collaborators. Especially to our collaborators in curiosity studies: Arjun Shankar, Peter T. Struck, David M. Lydon-Staley, and Dale Zhou. And to the colleagues who have modeled an expansive, de-disciplined curiosity by which we saw the world differently: Andrew Dilts, Andrea Pitts, Rachel & Jenn, Christopher Macklin, Edward T. Bullmore, and Mason A. Porter. Rambling curiously with you in the very heart of our fields and out past their edges has been an unspeakable pleasure.

We wish to thank Bob Prior at the MIT Press for securing the project and guiding it through to publication. We are so grateful to our gracious reviewers and readers, including Lynn Borton (of the *Choose to Be Curious* podcast), Sophia U. David, and Dale Zhou, and to our keen indexer, Mark Hineline. We are deeply honored to have worked with Poonam Mistry and are in awe of her creative artistry. And we offer special appreciation to the

Center for Curiosity, whose support provided much of the time and space to complete this work.

We offer profound thanks to our mother for safeguarding the magic of questioning things large and small. To our grandmother, for being our own personal Ole Worm. And to Silas, Simi, and Remy, who are indeed great big bundles of curiosity. Special thanks to Asia for so generously cultivating the space in which Perry could find these words.

And most of all, we are grateful for each other. Let's do this again sometime.

Notes

PREFACE

1. Abraham Flexner, *The Usefulness of Useless Knowledge* (Princeton: Princeton University Press, 2017), 78.

2. T. S. Eliot, "The Love Song of J. Alfred Prufrock," in *The Poems of T. S. Eliot: Collected and Uncollected Poems* (New York: Farrar, Straus, and Giroux, 2018), 1:6.

INTRODUCTION

1. Cf. Susan Engel, *The Hungry Mind: The Origins of Curiosity in Childhood* (Cambridge, MA: Harvard University Press, 2015), 5. Engel opens her book with the paradigmatically curious act of eating a bug, while here we imagine curiosity itself behaving like a bug.

2. Cf. Virginia Woolf, "Why?" (1942), in *Collected Essays* (New York: Harcourt, 1966), 2:278–283. In this essay, she wonderfully lampoons the practice of university lectures.

3. Virginia Woolf, *A Room of One's Own* (New York: Harcourt Brace Jovanovich, 1929), 5.

4. Woolf, *A Room of One's Own*, 4–8.

5. For more on the relationship between the sciences and humanities in our work together, see Perry Zurn and Danielle S. Bassett, "Seizing an Opportunity," *eLife Magazine* 8, no. e48336 (2019): 1–5.

6. Saint Augustine, *The Confessions* (Cambridge: Loeb Classical Library, 2014), especially Book II, Book X. See also Joseph Torchia, *Restless Mind: Curiositas and the Scope of Inquiry in St. Augustine's Psychology* (Milwaukee: Marquette University Press, 2013).

7. Francis Bacon, "Of Tribute" (1595), in *The Major Works* (Oxford: Oxford University Press, 2002), 31; Francis Bacon, *The Advancement of Learning* (London: Cassell and Company, 1893).

8. René Descartes, *The Passions of the Soul*, in *The Philosophical Writings of Descartes*, trans. John Cottingham, Robert Stoothoff, and Dugald Murdoch (1649; repr., Cambridge: Cambridge University Press, 1985), 1:325–404.

9. George Lowenstein, "The Psychology of Curiosity: A Review and Reinterpretation," *Psychological Bulletin* 116, no. 1 (1994): 75.

10. Daniel Berlyne, *Conflict, Arousal, and Curiosity* (Eastford, CT: Martino Fine Books, 1960).

11. Celeste Kidd and Benjamin Hayden, "The Psychology and Neuroscience of Curiosity," *Neuron* 88, no. 3 (2015): 449–460; Jacqueline Gottlieb, Michael Cohanpour, Yvonne Li, Nicholas Singletary, and Erfan Zabeh, "Curiosity, Information Demand, and Attentional Priority," *Current Opinion in Behavioral Sciences* 25 (2020): 83–91.

12. For a full-length study of the politics of curiosity, see Perry Zurn, *Curiosity and Power: The Politics of Inquiry* (Minneapolis: University of Minnesota Press, 2021).

13. For an unsurpassed model of this methodology in the field of anthropology, see Anna Tsing, *The Mushroom at the End of the World: On the Possibility of Life in Capitalist Ruins* (Princeton, NJ: Princeton University Press, 2015).

14. Ta-Nehisi Coates, *Between the World and Me* (New York: One World, 2015), 26.

15. Christina León, "Curious Entanglements: Opacity and Ethical Relation in Latina/o Aesthetics," in *Curiosity Studies: A New Ecology of Knowledge*, ed. Perry Zurn and Arjun Shankar (Minneapolis: University of Minnesota Press, 2020), 167–187; Erica Violet Lee, "'Be Safe, Nicimos': Indigenous Freedom and Curiosity in the Wastelands" (keynote speech at Carleton University, May 9, 2017), https://carleton.ca/circle/wp-content/uploads/CIR CLE-Conference-Keynote-Bio.pdf; Kristina Johnson, "Autism, Neurodiversity, and Curiosity," in *Curiosity Studies: A New Ecology of Knowledge*, ed. Perry Zurn and Arjun Shankar (Minneapolis: University of Minnesota Press, 2020), 129–146; Naoki Higashida, *The Reason I Jump* (New York: Random House, 2016).

16. Bas Hofstra, Vivek V. Kulkarni, Sebastian Munoz-Najar Galvez, Bryan He, Dan Jurafsky, and Daniel A. McFarland, "The Diversity-Innovation Paradox in Science," *Proceedings of the National Academy of the Sciences* 117, no. 17 (2020): 9284–9291; Rodrick Ferguson, *The Reorder of Things: The University and Its Pedagogies of Minority Difference* (Minneapolis: University of Minnesota Press, 2012).

17. Gilles Deleuze and Félix Guattari, *A Thousand Plateaus: Capitalism and Schizophrenia*, trans. Brian Massumi (Minneapolis: University of Minnesota Press, 1987).

CHAPTER 1

1. Joseph Glanvill, *Plus Ultra* (London: James Collins, 1668).

2. Benjamin Franklin, *The Autobiography of Benjamin Franklin* (London: J. Parsons, 1771).

3. Lorraine Daston, "On Scientific Observation," *Isis* 99, no. 1 (2008): 97–110, https://doi .org/10.1086/587535.

4. Aristotle, *On the Parts of Animals I–4*, trans. and ed. James G. Lennox (Oxford: Clarendon Press, 2002).

5. Sir Francis Bacon, *Novum Organum, or True Suggestions for the Interpretation of Nature* (1620; repr., New York: P. F. Collier, 1902). This openness is, not incidentally, deeply curtailed by colonialism, for which the ship at sea can also be a cipher.

6. Francis Bacon, *Instauratio Magna* (London: J. Bill, 1620), frontispiece. See British Museum number 1868,0808.3213, https://www.britishmuseum.org/collection/object /P_1868-0808-3213.

7. Robert Boyle, "Some Anatomical Observations of Milk Found in Veins, Instead of Blood; and of Grass, Found in the Wind-Pipes of Some Animals," *Philosophical Transactions of the Royal Society of London* 1, no. 6 (1665): 100–101; Robert Boyle, "An Account of a Very Odd Monstrous Calf," *Philosophical Transactions of the Royal Society of London* 1, no. 1 (1665): 10; unknown, "An Account of an Odd Spring in Westphalia, Together with an Information Touching Salt-Springs and the Straining of Salt-Water," *Philosophical Transactions of the Royal Society of London* 1, no. 7 (1665): 127–128.

8. Adiren Auzout, "Monsieur Auzout's Speculations of the Changes, Likely to Be Discovered in the Earth and Moon, by Their Respective Inhabitants," *Philosophical Transactions of the Royal Society of London* 1, no. 7 (1665): 120–123.

9. Ian Hasting, "Sir Francis Bacon and the Case for Curiosity," April 17, 2018, https://www .youtube.com/watch?v=I9zlM5cXTyE.

10. Abraham Flexner, *The Usefulness of Useless Knowledge* (Princeton, NJ: Princeton University Press, 2017), 53.

11. Flexner, *The Usefulness of Useless Knowledge*, 57.

12. Joseph Glanvill, *The Vanity of Dogmatizing* (London: Henry Eversden, 1661), https://www .exclassics.com/glanvil/glanvil.pdf.

13. T. C. Chamberlin, "The Method of Multiple Working Hypotheses," *Journal of Geology* 5 (1897): 755.

14. Victoria Skye, "Skye Blue Café Wall Illusion" (second prize in the Best Illusion of the Year Contest, 2017), http://illusionoftheyear.com/cat/top-10-finalists/2017/; Akiyoshi Kitaoka, "The Rice Wave Illusion," http://nautil.us/blog/ll/2685/2018-11-14+03:00:00; Corin A. Moss-Racusin, John F. Dovidio, Victoria L. Brescoll, Mark J. Graham, and Jo Handelsman, "Science Faculty's Subtle Gender Biases Favor Male Students," *Proceedings of the National Academy of Sciences USA* 109, no. 41 (2012): 16474–16479; Frances Trix and Carolyn Psenka, "Exploring the Color of Glass: Letters of Recommendation for Female and Male Medical Faculty," *Discourse and Society* 14 (2003): 191–220; Danielle Gaucher, Justin Friesen, and Aaron C. Kay, "Evidence that Gendered Wording in Job Advertisements Exists and Sustains Gender Inequality," *Journal of Personality and Social Psychology* 101 (2011): 109–128; Jazmin L. Brown-Iannuzzi, Brian Keith Payne, and Sophie Trawler, "Narrow Imaginations: How Imagining Ideal Employees Can Increase Racial Bias," *Group Processes and Intergroup Relations* 16 (2012): 661–670; Markus Helmer, Manuel Schottdorf, Andreas Neef, and Demian Battaglia, "Gender Bias in Scholarly Peer Review," *Elife* 6 (2017): e21718.

15. John R. Platt, "Strong Inference," *Science* 146, no. 3642 (1964): 347–353.

16. Aristotle, *Physics* (Cambridge, MA: Loeb Classical Library, 1957), VI:9, 239b15.

17. Stephen Graham, *The Gentle Art of Tramping* (1927; London: Bloomsbury Reader, 2019), 74.

18. Alonzo Reed and Brainerd Kellogg, *Graded Lessons in English: An Elementary English Grammar* (New York: Maynard, Merrill, and Co., 1896).

19. Angela Potochnik, "Levels of Explanation Reconceived," *Philosophy of Science* 77, no. 1 (2010): 59–72.

20. Philip W. Anderson, "More Is Different," *Science* 177, no. 4047 (1972): 393.

21. Arturo Rosenblueth and Norbert Wiener, "The Role of Models in Science," *Philosophy of Science* 12, no. 4 (1945): 320.

22. David M. Lydon-Staley, Perry Zurn, and Danielle S. Bassett, "Within-Person Variability in Curiosity during Daily Life and Associations with Well-Being," *Journal of Personality* 88, no. 4 (2020): 625–641; Alexandra Drake, Bruce P. Dore, Emily B. Falk, Perry Zurn, Danielle S. Bassett, and David M. Lydon-Staley, "Daily Stressor-Related Depressed Mood and Its Associations with Flourishing and Daily Curiosity," *Journal of Happiness Studies* (2021).

23. Helga Nowotny, *Insatiable Curiosity: Innovation in a Fragile Future* (Cambridge, MA: MIT Press, 2008); Perry Zurn, *Curiosity and Power: The Politics of Inquiry* (Minneapolis: University of Minnesota Press, 2021).

24. Flexner, *The Usefulness of Useless Knowledge*, 76.

25. How colonialism informed the spirit and manifestation of Ole Worm's collection—and therefore how imperial culture informed his curiosity—is a subject revisited in a later chapter.

26. Gottfried Wilhelm Leibniz, "Towards a Universal Characteristic" (1677), in *Leibniz Selections* (New York: Charles Scribner's Sons, 1979), 20.

27. Daniel E. Berlyne, "A Theory of Human Curiosity," *British Journal of Psychology* 45, no. 3 (1954): 180.

28. Virginia Woolf, "Hours in a Library," in *Collected Essays* (New York: Harcourt, 1925), 2:34.

29. Sylvain Bromberger, "Why-Questions," in *On What We Know We Don't Know: Explanation, Theory, Linguistics, and How Questions Shape Them* (Chicago: University of Chicago Press, 1993).

30. Tim Ingold, *The Life of Lines* (New York: Routledge, 2015), 66, figure 13.1.

31. George Loewenstein, "The Psychology of Curiosity: A Review and Reinterpretation," *Psychological Bulletin* 116, no. 1 (1994): 75.

32. David Andrich, "A Rating Formulation for Ordered Response Categories," *Psychometrika* 43, no. 4 (1978): 561–573.

33. Eunseong Cho, "Making Reliability Reliable: A Systematic Approach to Reliability Coefficients," *Organizational Research Methods* 19, no. 4 (2016): 651–682; Lee J. Cronbach, "Citation Classics," *Current Contents* 13 (1978): 263; Lee J. Cronbach, "Coefficient Alpha and the Internal Structure of Tests," *Psychometrika* 16, no. 3 (1951): 297–334.

34. Lydon-Staley, Zurn, and Bassett, "Within-Person Variability in Curiosity."

35. Jordan A. Litman, "Interest and Deprivation Factors of Epistemic Curiosity," *Personality and Individual Differences* 44, no. 7 (2008): 1585–1595; Todd B. Kashdan, Melissa C. Stiksma, David J. Disabato, Patrick E. McKnight, John Bekier, Joel Kaji, and Rachel Lazarus, "The Five-Dimensional Curiosity Scale: Capturing the Bandwidth of Curiosity and Identifying Four Unique Subgroups of Curious People," *Journal of Research in Personality* 73 (2018): 130–149.

36. Lauren N. Ross, "Causal Selection and the Pathway Concept," *Philosophy of Science* 85, no. 4 (2018): 551–572; Lauren N. Ross, "Causal Concepts in Biology: How Pathways Differ from Mechanisms and Why It Matters," *British Journal for the Philosophy of Science* 72, no. 1 (2018): 131–158; Carl Craver, "Mechanisms in Science," *Stanford Encyclopedia of Philosophy*, November 18, 2015, https://plato.stanford.edu/entries/science-mechanisms/.

37. Maxwell A. Bertolero and Danielle S. Bassett, "On the Nature of Explanations Offered by Network Science: A Perspective from and for Practicing Neuroscientists," *Topics in Cognitive Science* 12, no. 4 (2020): 1272–1293.

38. Virginia Woolf, "How Should One Read a Book?," *Collected Essays* (New York: Harcourt, 1925), 2:2.

39. Celeste Kidd and Benjamin Hayden, "The Psychology and Neuroscience of Curiosity," *Neuron* 88, no. 3 (2015): 449–460; Jacqueline Gottlieb, Pierres-Yves Oudeyer, Manuel Lopes, and Adrian Baranes, "Information-Seeking, Curiosity, and Attention: Computational and Neural Mechanisms," *Trends in Cognitive Sciences* 17, no. 11 (2013): 585–593.

40. Kent C. Berridge and Morten L. Kringelbach, "Pleasure Systems in the Brain," *Neuron* 86, no. 3 (2015): 646–664.

41. Kidd and Hayden, "The Psychology and Neuroscience of Curiosity."

42. Farshad Alizadeh Mansouri, Etienne Koechlin, Marcello G. P. Rosa, and Mark J. Buckley, "Managing Competing Goals—a Key Role for the Frontopolar Cortex," *Nature Reviews Neuroscience* 18 (2017): 645–657; David Badre, "Cognitive Control, Hierarchy, and the Rostro-Caudal Organization of the Frontal Lobes," *Trends in Cognitive Sciences* 12, no. 5 (2008): 193–200.

43. Mansouri et al., "Managing Competing Goals."

44. Gottlieb et al., "Information-Seeking, Curiosity, and Attention."

45. Loewenstein, "The Psychology of Curiosity," 79.

46. Robert Macfarlane, *Landmarks* (New York: Penguin Books, 2015).

47. Loewenstein, "The Psychology of Curiosity," 84.

48. Kidd and Hayden, "The Psychology and Neuroscience of Curiosity," 449.

49. Kidd and Hayden, "The Psychology and Neuroscience of Curiosity," 446.

50. Perry Zurn and Danielle S. Bassett, "Seizing an Opportunity," *Elife* 8 (2019): e48336.

CHAPTER 2

1. Edward Casey, *A World on Edge* (Bloomington: Indiana University Press, 2017), xv.

2. See, for example, Ilhan Inan, *The Philosophy of Curiosity* (New York: Routledge, 2012); Ilhan Inan, Lani Watson, Dennis Whitcomb, and Safiye Yigit, eds., *The Moral Psychology of Curiosity* (Lanham, MD: Rowman and Littlefield, 2018); Marianna Papastephanou, ed., *Toward New Philosophical Explorations of the Epistemic Desire to Know: Just Curious about Curiosity* (Cambridge, UK: Cambridge Scholars Publishing, 2019); Perry Zurn, *Curiosity and Power: The Politics of Inquiry* (Minneapolis: University of Minnesota Press, 2021).

3. Gilles Deleuze and Félix Guattari, *What Is Philosophy?*, trans. Hugh Tomlinson and Graham Burchell (New York: Columbia University Press, 1994), 2.

4. In the process of coming to know, we are always both knowers and unknowers, navigating a world of knowns and unknowns, which themselves belong to knowledges (i.e., systems of things known) and unknowledges (i.e., mysterious networks of things not yet known). And just as important, that project of coming to know requires the practice of unknowing—loosening edges, uprooting nodes, and reconfiguring networks. While much of this book focuses on knowledge network building, we return to the necessary companion project of unbuilding (or "cracking") in the conclusion.

5. Plutarch, "On Being a Busybody," in *Moralia VI* (Cambridge, MA: Loeb Classical Library, 2005), 515d–519a.

6. Saint Augustine, *On Genesis*, trans. Roland J. Teske (Washington, DC: Catholic University American Press, 1991), II.18.122.

7. Saint Augustine, *The Confessions*, trans. Carolyn J. B. Hammond (Cambridge, MA: Loeb Classical Library, 2014), X.35.211.

8. René Descartes, *The Passions of the Soul*, in *The Philosophical Writings of René Descartes,* trans. John Cottingham, Robert Stoothoff, and Dugald Murdoch (1649; repr., Cambridge: Cambridge University Press, 1985), §88.

9. John Locke, *Some Thoughts concerning Education* (1693; repr., Mineola, NY: Dover Philosophical Classics, 2013), 119.1, 120.3.

10. David Hume, *A Treatise on Human Nature* (1738; repr., Oxford: Oxford University Press, 2006), 2.3.10.1.

11. Inan, *The Philosophy of Curiosity*.

12. Ilhan Inan, "Curiosity, Truth, and Knowledge," in *The Moral Psychology of Curiosity*, ed. Ilhan Inan, Lani Watson, Dennis Whitcomb, and Safiye Yigit (Lanham, MD: Rowman and Littlefield, 2018), 18.

13. Lani Watson, "Educating for Curiosity," in *The Moral Psychology of Curiosity*, ed. Ilhan Inan, Lani Watson, Dennis Whitcomb, and Safiye Yigit (Lanham, MD: Rowman and Littlefield, 2018), 295–297; Lani Watson, "Curiosity and Inquisitiveness," in *The Routledge Handbook of Virtue Epistemology*, ed. Heather Battaly (New York: Routledge, 2018), 157.

14. Francois Joseph Gall, *On the Origin of the Moral Qualities and Intellectual Faculties of Man*, trans. Winslow Lewis (Boston: Marsh, Capen and Lyon, 1835), 115–120.

15. Johann Gaspar Spurzheim, *Outlines of Phrenology; Being Also a Manual of Reference for the Marked Busts* (London: Bagster and Thoms, 1827); H. Lundie, *The Phrenological Mirror; or, Delineation Book* (Leeds: C. Croshaw, 1844); George Combe, *The Constitution of Man in Relation to External Objects* (London: W. S. Orr and Co., 1847).

16. Danielle S. Bassett, Nicholas F. Wymbs, Mason A. Porter, Peter J. Mucha, Jean M. Carlson, and Scott T. Grafton, "Dynamic Reconfiguration of Human Brain Networks during Learning," *Proceedings of the National Academy of the Sciences* 108, no. 18 (2011): 7641–7646.

17. Celeste Kidd and Benjamin Hayden, "The Psychology and Neuroscience of Curiosity," *Neuron* 88, no. 3 (2014): 449–460; Jacqueline Gottlieb and Pierre Yves Oudeyer, "Towards a Neuroscience of Active Sampling and Curiosity," *Nature Reviews Neuroscience* 19 (2018): 758–770.

18. Perry Zurn, "Busybody, Hunter, Dancer: Three Historical Models of Curiosity," in *Toward New Philosophical Explorations of the Epistemic Desire to Know: Just Curious about Curiosity*, ed. Marianna Papastephanou (Cambridge, UK: Cambridge Scholars Publishing, 2019), 26–49.

19. Blaise Pascal, *Pensées* (1670), trans. A. J. Krailsheimer (New York: Penguin Books, 1995), §919–553.

20. Plutarch, "On Talkativeness," in *Moralia*, trans. W. C. Helmbold, vol. 6 (Cambridge, MA: Harvard University Press, 1939).

21. Michel Foucault, *The Hermeneutics of the Subject: Lectures at the College de France 1981–1982*, trans. Graham Burchell (New York: Picador, 2001), 222–223.

22. Juan Luis Vives, *On Education* (1531; repr., Cambridge: Cambridge University Press, 1913), 16.

23. Thomas Hobbes, *Leviathan* (1651; repr., Indianapolis: Hackett Publishers, 1994), 3.9.

24. Jean-Jacques Rousseau, *Émile, or Treatise on Education*, trans. William H. Payne (1762; repr., New York: D. Appleton and Company, 1909), 135.

25. Friedrich Nietzsche, *The Gay Science*, trans. Walter Kaufmann (New York: Vintage Books, 1974), §345, 283–284.

26. Rousseau, *Émile*, 135.

27. William James, *The Principles of Psychology* (1890; repr., New York: Dover Publications, 1918), 2:56.

28. William James, *Talk to Teachers on Psychology and to Students on Some of Life's Ideals* (1899; repr., New York: Dover Publications, 2001), 25; Charles Sanders Peirce, "Philosophy of

Mind," in *Collected Papers of Charles Sanders Peirce* (Cambridge, MA: Harvard University Press, 1966), 7:III.§7.509, 306.

29. James, *Talk to Teachers on Psychology*, 25.

30. John Dewey, *How We Think* (ReadaClassic, 2010), 26; John Dewey, *Democracy and Education* (New York: Simon and Brown, 2011), 115–116.

31. Michel Foucault, *The Use of Pleasure* (New York: Vintage Books, 1984).

32. Michel Foucault, "The Masked Philosopher" (1981), in *Ethics: Subjectivity and Truth*, ed. James Faubion (New York: New Press, 1998), 325–326.

33. Foucault, "The Masked Philosopher," 325–326.

34. Michel Foucault, "Friendship as a Way of Life" (1981), in *Ethics: Subjectivity and Truth*, ed. James Faubion (New York: New Press, 1998), 136–138.

35. Robin Wall Kimmerer, *Braiding Sweetgrass: Indigenous Wisdom, Scientific Knowledge, and the Teaching of Plants* (Minneapolis: Milkweed Editions, 2015), 53.

36. Kimmerer, *Braiding Sweetgrass*, 39.

37. Robin Wall Kimmerer, *Gathering Moss: A Natural and Cultural History of Mosses* (Corvallis: Oregon State University Press, 2003), 13.

38. Carl Mika, *Indigenous Education and the Metaphysics of Presence: A Worlded Philosophy* (New York: Routledge, 2017), 4.

39. Brian Burkhart, *Indigenizing Philosophy through the Land: A Trickster Methodology for Decolonizing Environmental Ethics and Indigenous Futures* (East Lansing: Michigan State University Press, 2019), xxviii.

40. Thomas Norton-Smith, *The Dance of Person and Place: One Interpretation of American Indian Philosophy* (Albany: SUNY Press, 2010), 55.

41. Henry James, "The Art of Fiction," in *The Future of the Novel: Essays on the Art of Fiction* (New York: Vintage Books, 1956), 12.

42. Desiderius Erasmus, *In Praise of Folly*, trans. Clarence Miller (1511; repr., New Haven, CT: Yale University Press, 2003), 51.

43. Adam Smith, "The History of Astronomy," *Glasgow Edition of the Works and Correspondence of Adam Smith* (Indianapolis: Liberty Fund, 1985), 3:1981.

CHAPTER 3

1. Ludwig van Beethoven, *Conversation-Book* (1820; repr., Suffolk, UK: Boydell Press, 2018).

2. Johann Wolfgang von Goethe, introduction to *Theory of Colours* (1810; repr., Cambridge, MA: MIT Press, 1970), lviii; 283, part V. 716.

3. Robert Macfarlane, *Landmarks* (New York: Penguin Books, 2015), 101.

4. Mary Carruthers, *The Craft of Thought: Meditation, Rhetoric, and the Making of Images, 400–1200* (Cambridge: Cambridge University Press, 1998), 77.

5. Béla Bollobás, *Modern Graph Theory* (New York: Springer, 2002).

6. Reproduced in Jeffrey F. Hamburger, *Script as Image* (Louvain: Peeters Publishers, 2014), 50.

7. Vatican library, Cod. Bibl. Chig. A. V. 129 I II, f. lr, reproduced in Ulrich Ernst, *Carmen Figuratum* (Vienna: Bohlau, 1991), 316.

8. Joaquin Goñi, Martijn P. van den Heuvel, Andrea Avena-Koenigsberger, Nieves Velez de Mendizabal, Richard F. Betzel, Alessandra Griffa, Patric Hagmann, et al., "Resting-Brain Functional Connectivity Predicted by Analytic Measures of Network Communication," *Proceedings of the National Academy of the Sciences* 111, no. 2 (2014): 833–838.

9. Victoria Finlay, *Color: A Natural History of the Palette* (London: Folio Society, 2009), 230. See also Yutaka Mino and Katherine R. Tsiang's translation, "skillfully molded like a full moon dyed with spring water," from the poem "The Secret-Colored Bowl Remaining from the Gifts to the Emperor," in *Ice and Green Clouds: Traditions of Chinese Celadon* (Indianapolis: Indianapolis Museum of Art in cooperation with Indiana University Press, 1986), 13; cp. Suzanne Staubach's translation, "The cups intriguingly fashioned like the full moon softened by spring water," from the poem "Secret Color Porcelain Teacup, a Leftover Tribute Piece," in *Clay: The History and Evolution of Humankind's Relationship with Earth's Most Primal Element* (2005; repr., Hanover, NH: University Press of New England, 2013), 75.

10. Edward Bullmore and Olaf Sporns, "The Economy of Brain Network Organization," *Nature Reviews Neuroscience* 13, no. 5 (2012): 336–349.

11. Nan Shepherd, "Fawn," in *In the Cairngorms and Other Poems* (Cambridge, UK: Galileo Publishers, 1934), 22.

12. Nan Shepherd, "Winter Branches," in *In the Cairngorms and Other Poems* (Cambridge, UK: Galileo Publishers, 1934), 35.

13. J. Watts Duncan and Steven H. Strogatz, "Collective Dynamics of 'Small-World' Networks," *Nature* 393, no. 6684 (1998): 440–442; Filippo Radicchi and Claudio Castellano, "Fundamental Difference between Superblockers and Superspreaders in Networks," *Physics Review* E 5, no. 1 (2017): 012318; Sandra González-Bailón, Javier Borge-Holthoefer, Alejandro Rivero, and Yamir Moreno, "The Dynamics of Protest Recruitment through an Online Network," *Scientific Reports* 1 (2011): 197.

14. Carter T. Butts, "Revisiting the Foundations of Network Analysis," *Science* 325, no. 5939 (2009): 414–416.

15. Richard F. Betzel and Danielle S. Bassett, "Multi-Scale Brain Networks," *Neuroimage* 160 (2017): 73–83.

16. Korbinian Brodmann, *Vergleichende Lokalisationslehre der Grosshirnrinde* (Leipzig: Johann Ambrosius Barth, 1909); Matthew F. Glasser, Timothy S. Coalson, Emma C. Robinson,

Carl D. Hacker, John Harwell, Essa Yacoub, Kamil Ugurbil, et al., "A Multi-Modal Parcellation of Human Cerebral Cortex," *Nature* 536, no. 7615 (2016): 171–178.

17. Edward T. Bullmore and Danielle S. Bassett, "Brain Graphs: Graphical Models of the Human Brain Connectome," *Annual Review of Clinical Psychology* 7 (2011): 113–140.

18. Roger E. Beaty, Scott Barry Kaufman, Mathias Benedek, Rex E. Jung, Yoed N. Kenett, Emanuel Jaulk, and Aljoscha C. Neubauer, "Personality and Complex Brain Networks: The Role of Openness to Experience in Default Network Efficiency," *Human Brain Mapping* 37, no. 2 (2016): 773–779; Roger E. Beaty, Mathias Benek, Scott Barry Kaufman, and Paul J. Silvia, "Default and Executive Network Coupling Supports Creative Idea Production," *Scientific Reports* 5 (2015): 10964; Aaron M. Scherer, Bradley C. Taber-Thomas, and Daniel Tranel, "A Neuropsychological Investigation of Decisional Certainty," *Neuropsychologia* 70 (2015): 206–213.

19. Nicholson Baker, *The Size of Thoughts: Essays and Other Lumber* (New York: Vintage Books, 1997), 24.

20. *Oxford English Dictionary*, oxforddictionaries.com; Daniel Webster, *American Dictionary of the English Language* (1828; repr., Chesapeake, VA: American Christian Education, 1967).

21. Maya Brainard, Frank Guenther, Jason Tourville, Alfonso Nieto-Castanon, Jean-Luc Anton, Bruno Nazarian, and F. Xavier Alario, "Distinct Representations of Phonemes, Syllables, and Supra-Syllabic Sequences in the Speech Production Network," *Neuroimage* 50, no. 2 (2010): 626–638.

22. W. Tecumseh Fitch and Angela D. Friederici, "Artificial Grammar Learning Meets Formal Language Theory: An Overview," *Philosophical Transactions of the Royal Society of London B: Biological Sciences* 367, no. 1598 (2012): 1933–1955.

23. Paul A. Luce and David B. Pisoni, "Recognizing Spoken Words: The Neighborhood Activation Model," *Ear and Hearing* 19, no. 1 (1998): 1–36.

24. Michael S. Vitevitch, "What Can Graph Theory Tell Us about Word Learning and Lexical Retrieval?," *Journal of Speech, Language, and Hearing Research* 51, no. 2 (2008): 408–422.

25. John F. Sowa, "Semantic Networks," in *Encyclopedia of Artificial Intelligence*, ed. Stuart C. Shapiro (New York: Wiley-Interscience, 1987).

26. Allan M. Collins and M. Ross Quillian, "Retrieval Time from Semantic Memory," *Journal of Verbal Learning and Verbal Behavior* 8 (1969): 240–248.

27. Frank C. Keil, *Semantic and Conceptual Development: An Ontological Perspective* (Cambridge, MA: Harvard University Press, 1979); Dan I. Slobin, "Cognitive Prerequisites for the Acquisition of Grammar," in *Studies of Child Language Development*, ed. Charles A. Ferguson and Dan I. Slobin (New York: Holt, Rinehart and Winston, 1973), 173–208.

28. Michael E. Bales and Stephen B. Johnson, "Graph Theoretic Modeling of Large-Scale Semantic Networks," *Journal of Biomedical Informatics* 39, no. 4 (2006): 451–464; Mark Steyvers and Joshua B. Tenenbaum, "The Large-Scale Structure of Semantic Networks:

Statistical Analyses and a Model of Semantic Growth," *Cognitive Science* 29, no. 1 (2005): 41–78.

29. John R. Anderson, *Learning and Memory: An Integrated Approach* (New York: Wiley, 2000).

30. David Danks, *Unifying the Mind: Cognitive Representations as Graphical Models* (Cambridge, MA: MIT Press, 2014); Sebastien Harispe, Sylvie Ranwez, Stefan Janaqi, and Jacky Montmain, "Semantic Similarity from Natural Language and Ontology Analysis," *Synthesis Lectures on Human Language Technologies* 8, no. 1 (2015): 1–254.

31. Yoed N. Kenett, Dror Y. Kenett, Eshel Ben-Jacob, and Miriam Faust, "Global and Local Features of Semantic Networks: Evidence from the Hebrew Mental Lexicon," *PLoS ONE* 6, no. 8 (2011): e23912.

32. T. H. White, *Mistress Masham's Repose* (London: Folio Society, 1989), 95.

33. Virginia Woolf, "Jane Austen," in *Collected Essays* (New York: Harcourt, 1967), 1:145; Virginia Woolf, *A Room of One's Own* (New York: Harcourt, 2005), 74.

34. Woolf, *A Room of One's Own*, 100.

35. Woolf, *A Room of One's Own*, 79.

36. Virginia Woolf, "The Historian and 'The Gibbon,'" in *Collected Essays* (New York: Harcourt, 1967), 1:115; Virginia Woolf, "The Pastons and Chaucer," in *Collected Essays* (New York: Harcourt, 1967), 3:15.

37. Virginia Woolf, "The Man at the Gate," in *Collected Essays* (New York: Harcourt, 1967), 3:217.

38. Virginia Woolf, "On Not Knowing Greek," in *Collected Essays* (New York: Harcourt, 1967), 1:11; Virginia Woolf, "Phases of Fiction," in *Collected Essays* (New York: Harcourt, 1967), 2:90.

39. Woolf, "Phases of Fiction," 2:90.

40. G. Stanley Hall and Theodate L. Smith, "Curiosity and Interest," *Pedagogical Seminary* 10 (1903): 315–358; William James, *Talks to Teachers on Psychology: And to Students on Some of Life's Ideals* (New York: Henry Holt, 1899).

41. Michael Taylor and Angeles I. Diaz, "On the Deduction of Galactic Abundances with Evolutionary Neural Networks," in *From Stars to Galaxies: Building the Pieces to Build Up the Universe*, ed. Antonella Vallenari, Rosaria Tantalo, Laura Portinari, and Alessia Moretti (San Francisco: Astronomical Society of the Pacific, 2008), 374:105–110.

42. Teresa M. Amabile, Karl G. Hill, Beth A. Hennessey, and Elizabeth M. Tighe, "The Work Preference Inventory: Assessing Intrinsic and Extrinsic Motivational Orientations," *Journal of Personality and Social Psychology* 66, no. 5 (1994): 950–967; Angelina R. Sutin, Lori L. Beason-Held, Susan M. Resnick, and Paul T. Costa, "Sex Differences in Resting-State Neural Correlates of Openness to Experience among Older Adults," *Cerebral Cortex* 19, no. 12 (2009): 2797–2802; Josephine D. Arasteh, "Creativity and Related Processes in the Young Child: A Review of the Literature," *Journal of Genetic Psychology* 112 (1968): 77–108.

43. Reka Z. Albert and Albert L. Barabási, "Statistical Mechanics of Complex Networks," *Reviews of Modern Physics* 74 (2002): 47–97.

44. Mark Steyvers and Joshua Tenenbaum, "The Large-scale Structure of Semantic Networks: Statistical Analyses and a Model of Semantic Growth," *Cognitive Sciences* 29, no. 1 (2005): 41–78.

45. Thomas T. Hills, Mounir Maouene, Josita Maouene, Adam Sheya, and Linda Smith, "Longitudinal Analysis of Early Semantic Networks: Preferential Attachment or Preferential Acquisition?," *Psychological Science* 20, no. 6 (2009): 729–739.

46. Yoed N. Kenett, David Anaki, and Mariam Faust, "Investigating the Structure of Semantic Networks in Low and High Creative Persons," *Frontiers in Human Neuroscience* 8 (2014): 407.

47. Danielle S. Bassett and Edward T. Bullmore, "Small-World Brain Networks Revisited," *Neuroscientist* 23, no. 5 (2017): 499–516.

48. Mark E. Newman, "Assortative Mixing in Networks," *Physical Review Letters* 89, no. 20 (2002): 208701.

49. Henrik Ronellenfitsch and Eleni Katifori, "Global Optimization, Local Adaptation, and the Role of Growth in Distribution Networks," *Physical Review Letters* 117, no. 13 (2016): 138301.

CHAPTER 4

1. We track the word "curiosity" through, for example, *polypragmosyne* (Greek), *periergia* (Greek), *curiositas* (Latin), *curiosité* (French), *Neugier* (German), *curiosity* (English), and so on.

2. Of course, as philosopher Roman Frigg would argue, literary figures and theoretical models have far more in common than one might think. See, for example, Roman Frigg, "Models and Fiction," *Synthese* 172 (2010): 251–268.

3. Stephen Hartmann and Roman Frigg, "Scientific Models," in *The Philosophy of Science: An Encyclopedia*, ed. Sahotra Sarkar and Jessica Pfeifer (New York: Routledge, 2005), 2:740–749.

4. Matthew Leigh, *From Polypragmon to Curiosus: Ancient Concepts of Curious and Meddlesome Behavior* (Oxford: Oxford University Press, 2013), 16–53; Jason van Niekerk, "The Virtue of Gossip," *South African Journal of Philosophy* 27, no. 4 (2013): 400–416.

5. Philo Judaeus, "De Agricultura," in *The Works of Philo Judaeus*, trans. Charles Duke Yonge (London: H. G. Bohn, 1854–1890), §34.

6. Saint Augustine, *Confessions*, trans. William Watts (Cambridge, MA: Harvard University Press, 2000), X.35.

7. Blaise Pascal, *Pensées* (1670), trans. A. J. Krailsheimer (New York: Penguin Books, 2003), §152.

8. Plutarch, "On Being a Busybody," in *Moralia VI*, trans. W. C. Helmbold (Cambridge, MA: Harvard University Press, 2005), 515–519.

9. Martin Heidegger, *Being and Time*, trans. Joan Stambaugh (New York: SUNY Press, 1996), 158–165.

10. Georges Bataille, *The Accursed Share*, trans. Robert Hurley (New York: Zone Books, 1991).

11. See van Niekerk, "The Virtue of Gossip."

12. See Perry Zurn, "Publicity and Politics: Foucault, the Prisons Information Group, and the Press," *Radical Philosophy Review* 17, no. 2 (2014): 94–105.

13. The word *explore* stems from the Latin verb *plorare*, meaning to cry out, wail, or lament. As ancient authors attest, the earliest sense of *explorare* was to reconnoiter a new hunting ground for game by shouting from different coordinates. This is an early connection between the art of hunting and the practice of curiosity.

14. Plutarch, "On Being a Busybody," 520e.

15. Augustine, *Confessions*, X.35.

16. Christian Zacher, *Curiosity and Pilgrimage: The Literature of Discovery in Fourteenth Century England* (Baltimore: Johns Hopkins University Press, 1976), chapter 4.

17. Richard de Bury, *Philobiblon*, ed. Samuel Hand, trans. John Bellingham Inglis (New York: Meyer Brothers and Company, 1899), chapter 8, 113–114. De Bury also speaks of enlisting underlings in the church to help grow his collections, calling them "the keenest of hunters [*venatores*]" (chapter 8, 120–121).

18. Friedrich Nietzsche, *Beyond Good and Evil* (1886), trans. Walter Kaufmann (New York: Random House, 1989), §45, 59.

19. Jacques Derrida, *The Animal that Therefore I Am*, ed. Marie-Louise Mallet, trans. David Wills (New York: Fordham University Press, 2008), 33, 14.

20. Zacher, *Curiosity and Pilgrimage*; Heidegger, *Being and Time*, 161.

21. Friedrich Nietzsche, *The Gay Science* (1882), trans. Walter Kaufmann (New York: Vintage Books, 1974), §375.

22. Friedrich Nietzsche, *Ecce Homo* (1908), in *On the Genealogy of Morals and Ecce Homo*, trans. Walter Kaufmann (New York: Vintage Books, 1989), §3.

23. Friedrich Nietzsche, *Twilight of the Idols* (1889), trans. R. J. Hollingdale (New York: Penguin Books, 1990), "Reason in Philosophy" §1.

24. Friedrich Nietzsche, *Thus Spoke Zarathustra* (1883), trans. Walter Kauffman (New York: Penguin Books, 1966), 41. This should be paired with his claim earlier that "God is dead" (12–13).

25. Nietzsche, *Twilight of the Idols*, "What the Germans Lack" §7.

26. Nietzsche, *The Gay Science*, §366. It is important to recognize that, while ableism has marked the uptake of Nietzsche's figure of the dancer, important work has been done to combat such. See, for example, CRIPSiE's film *New Constellation: A Dance-umentary* (2013), produced by Justin DuVal, Lindsay Eales, Danielle Peers, and Roxanne Ulanicki, https://vimeo.com/88104311.

27. Quoted in Scott Kaufman and Carolyn Gregoire, *Wired to Create: Unraveling the Mysteries of the Creative Mind* (New York: TarcherPerigee, 2015), 81.

28. Michel Foucault, *Ethics: Subjectivity and Truth*, ed. Paul Rabinow (New York: New Press, 1997), 325.

29. Friedrich Nietzsche, *Philosophy in the Tragic Age of the Greeks* (1873), trans. Marianne Cowen (Washington, DC: Regnery Publishing, 1996), xxx.

30. Helga Nowotny, *Insatiable Curiosity: Innovation in a Fragile Future* (Cambridge, MA: MIT Press, 2008).

31. See Perry Zurn, "Curiosities at War: The Police and Prison Resistance after May '68," *Modern and Contemporary France* 26, no. 2 (2018): 179–191; Perry Zurn, "The Curiosity at Work in Deconstruction," *Journal of French and Francophone Philosophy* 26, no. 1 (2018): 65–87; Perry Zurn, "Curiosity and Political Resistance," in *Curiosity Studies: Toward a New Ecology of Knowledge*, ed. Perry Zurn and Arjun Shankar (Minneapolis: University of Minnesota Press, 2020), 227–245; Perry Zurn, *Curiosity and Power: The Politics of Inquiry* (Minneapolis: University of Minnesota Press, 2021).

32. See Michael North, *Novelty: A History of the New* (Chicago: University of Chicago Press, 2013).

33. Gloria Anzaldúa, *Interviews/Entrevistas*, ed. AnaLouise Keating (New York: Routledge, 2000), 252; Gloria Anzaldúa, "La Prieta," in *This Bridge Called My Back*, ed. Cherríe Moraga and Gloria Anzaldúa (1981; repr., New York: SUNY Press, 2015), 202.

34. Anzaldúa, *Interviews/Entrevistas*, 25.

35. Gloria Anzaldúa, *Light in the Dark / Luz en lo Obscuro*, ed. AnaLouise Keating (Durham, NC: Duke University Press, 2015), 102, 103, 130.

36. Anzaldúa, *Light in the Dark*, 102, 103, 114. We can't help but hear the resonance with Virginia Woolf, who we previously found rolling "a sentence or two on my tongue" as if sucking a blackcurrant candy.

37. Anzaldúa, *Light in the Dark*, 82, 123, 83, 114.

CHAPTER 5

1. Johann Wolfgang von Goethe, *The Metamorphosis of Plants*, photographer Gordon L. Miller (Cambridge, MA: MIT Press, 2009).

2. These sociocultural processes are particularly evident in the growth of scientific knowledge, where the culture of science determines the value placed on each kind of question. See, for

example, Karine Chelma and Evelyn Fox Keller, *Cultures without Culturalism: The Making of Scientific Knowledge* (Durham, NC: Duke University Press, 2017). Unfortunately, that culture is created by humans, thereby making the vision of scientific inquiry somewhat less bright. Humans generally share explicit and implicit biases against people of marginalized genders, races, and ethnicities, and thus human scientific culture extols the questions of some questioners more than others. See, for example, Jordan D. Dworkin, Kristin A. Linn, Erin G. Teich, Perry Zurn, Russell T. Shinohara, and Danielle S. Bassett, "The Extent and Drivers of Gender Imbalance in Neuroscience Reference Lists," *Nature Neuroscience* 23, no. 8 (2020): 918–926; Paula Chakravartty, Rachel Kuo, Victoria Grubbs, and Charlton McIlwain, "#CommunicationSoWhite," *Journal of Communication* 68, no. 2 (2018): 254–266; Michelle L. Dion, Jane Lawrence Sumner, and Sara McLaughlin Mitchell, "Gendered Citation Patterns across Political Science and Social Science Methodology Fields," *Political Analysis* 26, no. 3 (2018): 312–327. The unknowns in science, both the local holes and the giant space surrounding the knowns, are therefore in part the answers to the questions posed by marginalized thinkers. See, for example, Bas Hofstra, Vivek V. Kulkarni, Sebastian Munoz-Najar Galvez, Bryan He, Dan Jurafsky, and Daniel A. McFarland, "The Diversity-Innovation Paradox in Science," *Proceedings of the National Academy of Sciences* 117, no. 17 (2020): 9284–9291. Our work is cut out for us in developing an equitable trajectory of scientific inquiry that charts our forays into the space of the unknowns in a way that freely searches for the truth, throwing off the chains that constrain us to the paths of an arbitrarily select few. See, for example, Jordan Dworkin, Perry Zurn, and Danielle S. Bassett, "(In)citing Action to Realize an Equitable Future," *Neuron* 106, no. 6 (2020): 890–894.

3. Maximilian Fürbringer, *Untersuchungen zur Morphologie und Systematik der Vögel* (Amsterdam: van Halkema, 1888).

4. Robert Siegfried, *From Elements to Atoms: A History of Chemical Composition* (Philadelphia: American Philosophical Society, 2002), 92.

5. Robert Oerter, *The Theory of Almost Everything: The Standard Model, the Unsung Triumph of Modern Physics* (New York: Penguin Group, 2006).

6. Benjamin Franklin, *The Autobiography of Benjamin Franklin* (London: J. Parsons, 1771).

7. Katy Börner, Olga Scrivner, Mike Gallant, Shutian Ma, Xiaozhong Liu, Keith Chewning, Lingfei Wu, and James A. Evans, "Skill Discrepancies between Research, Education, and Jobs Reveal the Critical Need to Supply Soft Skills for the Data Economy," *Proceedings of the National Academy of Sciences* 115, no. 50 (2018): 12630–12637.

8. Gottfried Wilhelm Leibniz, "Towards a Universal Characteristic" (1677), in *Leibniz Selections* (New York: Charles Scribner's Sons, 1979).

9. Elad Ganmor, Ronen Segev, and Elad Schneidman, "A Thesaurus for a Neural Population Code," *Elife* 8, no. 4 (September 2015): e06134.

10. Henri Poincaré, *Science and Hypothesis* (1902; repr., London: Walter Scott Publishing Co., 1914), xxiv.

11. Lorenz Oken, *Elements of Physiophilosophy* (London: Ray Society, 1847), 3.

12. John Dewey, *Democracy and Education* (London: Free Press, 1916), 340.

13. Ernst Haeckel, *Report on the Radiolaria Collected by H.M.S. Challenger during the Years 1873–1876* (Chapel Hill, NC: Gutenberg Project, 2013), plates.

14. Peter Bearman, James Moody, and Robert Faris, "Networks and History," *Complexity* 8, no. 1 (2003): 61–71.

15. Nicolas H. Christianson, Ann Sizemore Blevins, and Danielle S. Bassett, "Architecture and Evolution of Semantic Networks in Mathematics Texts," *Proceedings of the Royal Society A* 476 (2020): 20190741.

16. Ann Sizemore, Elisabeth Karuza, Chad Giusti, and Danielle S. Bassett, "Knowledge Gaps in the Early Growth of Semantic Feature Networks," *Nature Human Behavior* 2, no. 9 (2018): 682–692.

17. Walt Whitman, "On the Beach at Night Alone," in *Leaves of Grass* (Boston: James R. Osgood and Company, 1881–1882), 220.

18. David Lydon-Staley, Dale Zhou, Ann Sizemore Blevins, Perry Zurn, and Danielle S. Bassett, "Hunters, Busybodies, and the Knowledge Network Building Associated with Deprivation Curiosity," *Nature Human Behavior* 5 (2021): 327–336.

19. Perry Zurn, "Busybody, Hunter, Dancer: Three Historical Models of Curiosity," in *Toward New Philosophical Explorations of the Epistemic Desire to Know: Just Curious about Curiosity*, ed. Marianna Papastephanou (Newcastle upon Tyne: Cambridge Scholars, 2019), 26–49.

20. Friedrich Nietzsche, *Beyond Good and Evil* (1886), trans. Walter Kaufmann (New York: Random House, 1989), §45, 59; Philo Judaeus, "De Agricultura," in *The Works of Philo Judaeus*, trans. Charles Duke Yonge (London: H. G. Bohn, 1854–1890), §34.

21. Todd B. Kashdan, Melissa C. Stiksma, David J. Disabato, Patrick E. McKnight, John Bekier, Joel Kaji, and Rachel Lazarus, "The Five-Dimensional Curiosity Scale: Capturing the Bandwidth of Curiosity and Identifying Four Unique Subgroups of Curious People," *Journal of Research in Personality* 73 (2018): 130–149.

22. Robert Macfarlane, *Landmarks* (New York: Penguin Books, 2015).

23. Richard N. Aslin, "Statistical Learning: A Powerful Mechanism that Operates by Mere Exposure," *Wiley Interdisciplinary Reviews Cognitive Science* 8, no. 1–2 (2017): 10.1002/wcs.1373; Steven H. Tompson, Ari E. Kahn, Emily B. Falk, Jean M. Vettel, and Danielle S. Bassett, "Individual Differences in Learning Social and Nonsocial Network Structures," *Journal of Experimental Psychology: Learning, Memory, Cognition* 45, no. 2 (2019): 253–271.

24. Christopher W. Lynn, Lia Papadopoulos, Ari E. Kahn, and Danielle S. Bassett, "Human Information Processing in Complex Networks," *Nature Physics* 16 (2020): 965–973.

25. Elisabeth A. Karuza, Ari E. Kahn, Sharon L. Thompson-Schill, and Danielle S. Bassett, "Process Reveals Structure: How a Network Is Traversed Mediates Expectations about Its Architecture," *Scientific Reports* 7, no. 1 (2017): 12733; Anna C. Schapiro, Timothy

R. Rogers, Natalia I. Cordova, Nicholas B. Turk-Browne, and Matthew M. Botvinick, "Neural Representations of Events Arise from Temporal Community Structure," *Nature Neuroscience* 16, no. 4 (2013): 486–492; Ari E. Kahn, Elisabeth A. Karuza, Jean M. Vettel, and Danielle S. Bassett, "Network Constraints on Learnability of Probabilistic Motor Sequences," *Nature Human Behavior* 2, no. 12 (2018): 936–947.

26. Lynn et al., "Human Information Processing."

27. Christopher W. Lynn and Danielle S. Bassett, "How Humans Learn and Represent Networks," *Proceedings of the National Academy of Sciences* 117, no. 47 (2020): 29407–29415.

28. Arthur C. Benson, *Ruskin: A Study in Personality* (London: Smith, Elder and Co., 1911), 295.

29. William Morris, *News from Nowhere* (1890; repr., New York: Penguin Classics, 2012).

30. Virginia Woolf, "Why?," in *Collected Essays* (New York: Harcourt, 1953), 2:279–280.

31. W. Ross Ashby, "Principles of the Self-Organizing System," in *Principles of Self-Organization: Transactions of the University of Illinois Symposium*, ed. Heinz von Foerster and George W. Zopf Jr. (London: Pergamon Press: 1962), 255–278.

32. Christopher Lynn, Ari E. Kahn, Nathaniel Nyema, and Danielle S. Bassett, "Abstract Representations of Events Arise from Mental Errors in Learning and Memory," *Nature Communications* 11, no. 1 (2020): 2313.

33. Plato, *The Republic* (Cambridge, MA: Loeb Classical Library, 2013), 514a–520a.

34. Lynn et al., "Abstract Representations of Events."

35. Joseph Glanvill, *The Vanity of Dogmatizing* (1661; repr., New York: Columbia University Press, 1931), 30.

36. Lynn et al., "Human Information Processing."

37. Lynn et al., "Abstract Representations of Events."

38. Immanuel Kant, *Critique of Pure Reason*, trans. Norman Kemp Smith (New York: Palgrave Macmillan, 2016), 607–608.

39. Simon De Deyne, Danielle J. Navarro, Andrew Perfors, and Gert Storms, "Structure at Every Scale: A Semantic Network Account of the Similarities between Unrelated Concepts," *Journal of Experimental Psychology: General* 145, no. 9 (2016): 1228–1254.

40. Robert Macfarlane, Dan Richards, and Stanley Donwood, *Holloway* (London: Faber and Faber, 2013).

41. Mike Bostock, "Visualizing Algorithms," 2020, https://bost.ocks.org/mike/algorithms.

42. Douglas L. Nelson and Ningchuan Zhang, "The Ties that Bind What Is Known to the Recall of What Is New," *Psychonomic Bulletin and Review* 7, no. 4 (2000): 604–617; Michelle Lee, Gary E. Martin, Abigail Hogan, Deanna Hano, Peter C. Gordon, and Molly Losh, "What's the Story? A Computational Analysis of Narrative Competence in Autism," *Autism* (2017): 1362361316677957; Kelly Renz, Elizabeth Pugzles Lorch, Richard Milich,

Clarese Lemberger, Anna Bodner, and Richard Welsh, "On-Line Story Representation in Boys with Attention Deficit Hyperactivity Disorder," *Journal of Abnormal Child Psychology* 31, no. 1 (2003): 93–104; Claudia Drummond, Gabriel Coutinho, Rochele Paz Fonseca, Naima Assuncao, Alina Teldeschi, Ricardo de Oliveira-Souza, Jorge Moll, et al., "Deficits in Narrative Discourse Elicited by Visual Stimuli Are Already Present in Patients with Mild Cognitive Impairment," *Frontiers in Aging Neuroscience* 7 (2015): 96; Manfred Spitzer, "Associative Networks, Formal Thought Disorders and Schizophrenia: On the Experimental Psychopathology of Speech-Dependent Thought Processes," *Der Nervenarzt* 64, no. 3 (1993): 147–159.

43. Christianson, Sizemore Blevins, and Bassett, "Architecture and Evolution of Semantic Networks"; Henrique F. de Arruda, Luciano da F. Costa, and Diego R. Amancio, "Using Complex Networks for Text Classification: Discriminating Informative and Imaginative Documents," *Europhysics Letters* 113, no. 2 (2016): 28007; Lucy R. Chai, Dale Zhou, and Danielle S. Bassett, "Evolution of Semantic Networks in Biomedical Texts," *Journal of Complex Networks* 8, no. 1 (2020): cnz023.

44. Angeliki Lazaridou, Marco Marelli, and Marco Baroni, "Multimodal Word Meaning Induction from Minimal Exposure to Natural Text," *Cognitive Science* 41, no. S4 (2017): 677–705.

45. Tomas Mikolov, Kai Chen, Greg Corrado, and Jeffrey Dean, "Efficient Estimation of Word Representations in Vector Space," 2013, arXiv:1301.3781.

46. Zurn, "Busybody, Hunter, Dancer."

47. Andrea Pintos, Charlton Cheung, Simon De Deyne, Christy L. M. Hui, and Eric Y. H. Chen, "Detecting Semantic Distance Abnormalities in Psychosis: Quantification of Word Associations Using Semantic Space Modeling," *Schizophrenia Bulletin* 46 (2020): S252.

48. William Somerset Maugham, *The Summing Up* (New York: Doubleday, Doran and Company, 1938).

49. Alexandra N. M. Darmon, Marya Bazzi, Sam D. Howison, and Mason A. Porter, "Pull Out All the Stops: Textual Analys Analysis via Punctuation Sequences," 2018, https://arxiv.org/abs/1901.00519.

CHAPTER 6

1. Hans Baltussen, *The Peripatetics: Aristotle's Heirs 322 BCE—200 CE* (New York: Routledge, 2016).

2. Lauren N. Ross, "Causal Selection and the Pathway Concept," *Philosophy of Science* 85 (2018): 551–572.

3. Matthew Leigh, *From Polypragmon to Curiosus: Ancient Concepts of Curious and Meddlesome Behavior* (Oxford: Oxford University Press, 2013).

4. Christian Zacher, *Curiosity and Pilgrimage: The Literature of Discovery in Fourteenth Century England* (Baltimore: Johns Hopkins University Press, 1976).

5. Justin Stagl, *A History of Curiosity: The Theory of Travel 1550–1800* (Chur, Switzerland: Harwood Academic Publishers, 1995); Nigel Leask, *Curiosity and the Aesthetics of Travel Writing, 1770–1840* (Oxford: Oxford University Press, 2002).

6. Daniel Gade, *Curiosity, Inquiry, and the Geographical Imagination* (New York: Peter Lang, 2011).

7. Walter Benjamin, *The Arcades Project*, trans. Howard Eiland and Kevin McLaughlin (1927–1940; repr., Cambridge, MA: Harvard University Press, 1999), 81.

8. Jean-Jacques Rousseau, *Les Confessions* (Paris: Pléiade Édition, 1933), livre 4; Jean-Jacques Rousseau, *The Confessions* (London: Wordsworth Editions, 1996), 157.

9. Friedrich Nietzsche, *Thus Spoke Zarathustra* (1885; repr., New York: Penguin Books, 1954), 40–41.

10. Annabel Abbs, "In Search of the Body: Simone de Beauvoir," in *Windswept: Walking the Paths of Trailblazing Women* (New York: Tin House Books, 2021); Kate Kirkpatrick, *Becoming Beauvoir: A Life* (New York: Bloomsbury Academic, 2019).

11. Silvia Montiglio, *Wandering in Ancient Greek Culture* (Chicago: University of Chicago Press, 2005), esp. chapter 7.

12. Édouard Glissant, *Introduction à une poétique du divers* (Paris: Gallimard, 1996), 43–44. One might also think here of Cornel West's distinction between thinking in the mode of jazz improvisation versus thinking like a classical symphony. See Cornel West, "Cornel West: Truth," in *Examined Life: Excursions with Contemporary Thinkers*, ed. Astra Taylor (New York: New Press, 2009), 1–24.

13. Plato, *Apology*, in *Euthyphro, Apology, Crito, Phaedo, Phaedrus* (Cambridge, MA: Loeb Classical Library, 2017), 22a7.

14. Plato, *Greater Hippias*, in *Cratylus, Parmenides, Greater Hippias, Lesser Hippias* (Cambridge, MA: Loeb Classical Library, 1926), 304c1–304c2.

15. Plato, *Apology*, 30e.

16. Plato, *Phaedrus*, in *Euthyphro, Apology, Crito, Phaedo, Phaedrus* (Cambridge, MA: Loeb Classical Library, 2017), 251d1–251d8.

17. Montiglio, *Wandering in Ancient Greek Culture*, 153. Compare this to the following: "Wandering . . . drowns the Eleatic principles of identity and noncontradiction in the ocean that Parmenides refused to sail" (171).

18. Plato, *The Republic* (Cambridge, MA: Loeb Classical Library 2013), 484b.

19. The marked exception here is *The Laws*, which takes place entirely while walking.

20. Plato, *Theaetetus*, in *Theaetetus, Sophist* (Cambridge, MA: Loeb Classical Library, 1921), 150b–150c.

21. Plato, *Statesman*, in *Statesman, Philebus, Ion* (Cambridge, MA: Loeb Classical Library, 1925).

22. Astra Taylor, ed., *Examined Life: Excursions with Contemporary Thinkers* (New York: New Press, 2009).

23. Thich Nhat Hanh, *How to Walk* (Berkeley, CA: Parallax Press, 2015), 12, 27, 39, 62–63, 86, 112. See also Nguyen Anh-Huong and Thich Nhat Hanh, *Walking Meditation* (Louisville, KY: Sounds True, 2006).

24. Zacher, *Curiosity and Pilgrimage*, esp. chapter 3.

25. Robin Wall Kimmerer, *Braiding Sweetgrass: Indigenous Wisdom, Scientific Knowledge, and the Teachings of Plants* (Minneapolis: Milkweed Editions, 2013), 205–215. See also Leanne Betasamosake Simpson, *As We Have Always Done: Indigenous Freedom through Radical Resistance* (Minneapolis: University of Minnesota Press, 2017), 56–58; Edward Benton-Banai, *The Mishomis Book: The Voice of the Ojibway* (1988; repr. Minneapolis: University of Minnesota Press, 2010), 6–11.

26. Joseph Amato, *On Foot: A History of Walking* (New York: NYU Press, 2004), 105.

27. Rousseau, *Les Confessions*, 157.

28. Jean-Jacques Rousseau, *Reveries of a Solitary Walker* (New York: Penguin, 1980).

29. Frédéric Gros, *A Philosophy of Walking* (New York: Verso, 2014), 82.

30. Henry David Thoreau, *Walking* (1862; repr., Carlisle, MA: Applewood Books, 1992), 8, 11, 14, 28, 40, 53, 55.

31. Walt Whitman, "In Paths Untrodden," in *Leaves of Grass* (1892; repr., New York: Bantam Classics, 1983), 92.

32. Walt Whitman, "Song of the Open Road," in *Leaves of Grass* (1892; repr., New York: Bantam Classics, 1983), 121–127.

33. Walt Whitman, "Out from behind the Mask," in *Leaves of Grass* (1892; repr., New York: Bantam Classics, 1983), 307–308; Walt Whitman, "Thought," in *Leaves of Grass* (1892; repr., New York: Bantam Classics, 1983), 312.

34. Walt Whitman, "As I Walk These Broad Majestic Days," in *Leaves of Grass* (1892; repr., New York: Bantam Classics, 1983), 386–387.

35. Whitman, "Song of the Open Road," 122.

36. Mary Oliver, "How I Go to the Woods," in *Devotions: The Selected Poems of Mary Oliver* (New York: Penguin Press, 2017), 64.

37. Virginia Woolf, "Street Haunting: A London Adventure," in *Collected Essays* (1930; repr., New York: Harcourt, 1967), 4:156.

38. Annie Dillard, "Seeing," in *Pilgrim at Tinker Creek* (1974; repr., New York: HarperPerennial, 2013), 35.

39. In this regard, consider recent work on the possibility (and impossibility) of the Black *flâneur*. See, for example, Francesca Sobande, Alice Schoonejans, Guillaume D. Johnson,

Kevin D. Thomas, and Anthony Kwame Harrison, "Strolling with a Question: Is It Possible to Be a Black Flâneur?," Independent Social Research Foundation, May 21, 2021.

40. Gloria Anzaldúa, *Light in the Dark / Luz en lo Obscuro*, ed. AnaLouise Keating (Durham, NC: Duke University Press, 2015), 91–92, 101.

41. Guy Debord, *Society of the Spectacle* (1967; repr., Oakland, CA: AK Press 2018), §10.

42. Libero Andreotti and Xavier Costa, eds., *Theory of the Dérive and Other Situationist Writings* (Barcelona: Museu d'Art Contemporani de Barcelona, 1996), 40–41.

43. Guy Debord, "Theory of the Dérive," in *Theory of the Dérive and Other Situationist Writings*, ed. Libero Andreotti and Xavier Costa (Barcelona: Museu d'Art Contemporani de Barcelona, 1996), 22–26.

44. María Lugones, "Tactical Strategies of the Streetwalker / Estrategias Tacticas de la Callejera," in *Pilgrimages/Peregrinajes: Theorizing Coalition against Multiple Oppressions* (Lanham, MD: Rowman and Littlefield, 2003), 209–210, 220–221.

45. Brian B. Knudsen, Terry Nichols Clark, and Daniel Silver, "Walking, Social Movements, and Arts Activities in the United States, Canada, and France," in *Walking in Cities: Quotidian Mobility as Urban Theory, Method, and Practice* (Philadelphia: Temple University Press, 2015), 229–246.

46. Margaret Quintan and Benjamin Gates, "'Walking the City': Performance of Strategies and Tactics in the 1985 Bus Accessibility Protests," *Disability Studies Quarterly* 32, no. 1 (2012).

47. Tim Ingold, *The Life of Lines* (New York: Routledge, 2015), 49.

48. Tim Ingold, "Footprints through the Weather-World: Walking, Breathing, Knowing," in *Making Knowledge: Explorations of the Indissoluble Relation between Mind, Body, and Environment*, ed. Trevor H. J. Marchand (Oxford: Wiley-Blackwell, 2001), 121–122.

49. Tim Ingold, *Lines: A Brief History* (New York: Routledge, 2007), 89.

50. Ingold, *The Life of Lines*, 47.

51. Tim Ingold, "Ways of Mind-Walking: Reading, Writing, Painting," *Visual Studies* 25, no. 1 (2010): 15–23.

52. Michel de Certeau, "Walking in the City," in *The Practice of Everyday Life* (Berkeley: University of California Press, 1984), 98, 99, 101, 107.

CHAPTER 7

1. Percy Bysshe Shelley, "Hymn to Intellectual Beauty," in *Rosalind and Helen, A Modern Eclogue* (1876), 67.

2. Maxwell A. Bertolero and Danielle S. Bassett, "How the Mind Emerges from the Brain's Complex Networks," *Scientific American*, July 2019; Arthur W. Toga and John C. Mazziotta, eds., *Brain Mapping: The Systems* (Cambridge, MA: Academic Press, 2000).

3. Duncan J. Watts and Steven Strogatz, "Collective Dynamics of 'Small-World' Networks," *Nature* 393, no. 6684 (1998): 440–442; Stanley Milgram, "The Small World Problem," *Psychology Today* 1 (1967): 61–67.

4. Melanie Mitchell, *Complexity: A Guided Tour* (Oxford: Oxford University Press, 2011).

5. Danielle S. Bassett and Edward T. Bullmore, "Small-World Brain Networks Revisited," *Neuroscientist* 23, no. 5 (2017): 499–516; Danielle S. Bassett and Edward T. Bullmore, "Small-World Brain Networks," *Neuroscientist* 12, no. 6 (2006): 512–523.

6. Maxwell A. Bertolero and Danielle S. Bassett, "On the Nature of Explanations Offered by Network Science: A Perspective from and for Practicing Neuroscientists," *Topics in Cognitive Science* 12, no. 4 (2020): 1272–1293.

7. Danielle S. Bassett and Olaf Sporns, "Network Neuroscience," *Nature Neuroscience* 20, no. 3 (2017): 353–364; Danielle S. Bassett, Perry Zurn, and Joshua I. Gold, "On the Nature and Use of Models in Network Neuroscience," *Nature Reviews Neuroscience* 19, no. 9 (2018): 566–578.

8. Carl F. Craver, "Mechanisms in Science," in *Stanford Encyclopedia of Philosophy*, November 18, 2015, https://plato.stanford.edu/entries/science-mechanisms/#MecAcc.

9. Lauren Ross, "Causal Concepts in Biology: How Pathways Differ from Mechanisms and Why It Matters," *British Journal for the Philosophy of Science* 72, no. 1 (2021): 13–158.

10. Norton Juster, *The Phantom Tollbooth* (New York: Alfred A. Knopf, 1988).

11. Olaf Sporns and Richard Betzel, "Modular Brain Networks," *Annual Review of Psychology* 67 (2016): 613–640; Courtney Gallen and Mark D'Esposito, "Brain Modularity: A Biomarker of Intervention-Related Plasticity," *Trends in Cognitive Sciences* 23, no. 4 (2019): 293–304.

12. Marcus Raichle, "The Brain's Default Mode Network," *Annual Review of Neuroscience* 38 (July 2015): 433–447; Christopher G. Davey and Ben J. Harrison, "The Brain's Center of Gravity: How the Default Mode Network Helps Us to Understand the Self," *World Psychiatry* 17, no. 3 (2018): 278–279; Pengmin Qin, Simone Grimm, Niall W. Duncan, Yan Fan, Zirui Huang, Timothy Lane, Xuchu Weng, et al., "Spontaneous Activity in Default-Mode Network Predicts Ascription of Self-Relatedness to Stimuli," *Social Cognitive and Affective Neuroscience* 11, no. 4 (2016): 693–702.

13. Jerry Fodor, *The Modularity of Mind* (Cambridge, MA: MIT Press, 1983).

14. Marc Kirschner and John Gerhart, "Evolvability," *Proceedings of the National Academy of Sciences* 95, no. 15 (1998): 8420–8427; Gerard Schlosser and Gunter Wagner, eds., *Modularity in Development and Evolution* (Chicago: University of Chicago Press, 2004).

15. Celeste Kidd and Benjamin Hayden, "The Psychology and Neuroscience of Curiosity," *Neuron* 88, no. 3 (2015): 449–460.

16. Marieke Mur, Peter A. Bandettini, and Nikolaus Kriegeskorte, "Revealing Representational Content with Pattern-Information fMRI—an Introductory Guide," *Social Cognitive and Affective Neuroscience* 4, no. 1 (2009): 101–109.

17. Stefano Anzellotti and Marc N. Coutanche, "Beyond Functional Connectivity: Investigating Networks of Multivariate Representations," *Trends in Cognitive Sciences* 22, no. 3 (2018): 258–269; Marc N. Coutanche, Essang Akpan, and Rae R. Buckser, "Representational Connectivity Analysis: Identifying Networks of Shared Changes in Representational Strength through Jackknife Resampling," *bioRxiv* (2020), doi: https://doi.org/10.1101/2020.05.28.103077; Alessio Basti, Marieke Mur, Nikolaus Kriegeskorte, Vittorio Pizzella, Laura Marzetti, and Olaf Hauk, "Analysing Linear Multivariate Pattern Transformations in Neuroimaging Data," *PLoS ONE* 14, no. 10 (2019): e0223660.

18. John Considine, *Small Dictionaries and Curiosity* (Oxford: Oxford University Press, 2017), 1. The quote is a translation of Scaliger's poem originally written in Latin: Joseph Justus Scaliger, *Poemata* 35 (no. 39, "In lexicorum compilatores, inscriptum lexico Arabico a se collecto, in Batavis"), "Lexica contexat." John Considine, *Dictionaries in Early Modern Europe: Lexicography and the Early Making of Heritage* (Cambridge: Cambridge University Press, 2008). Considine notes that he draws this quotation from, for example, Franciscus Junius in Franciscus Junius and Thomas Marshall, eds., *Quatuor D.N. Jesu Christi Euangeliorum versiones perantiquae duae, Gothica scil. Et Anglo-Saxonica* (Dordrecht, 1665). "Si quem dura manet sententia Judicis, olim / Damnatum aerumnis suppliciisque caput: / Hunc neque fabrili lassent ergastula massa, / Nec rigidas vexent fossa metalla manus. / Lexica contexat. Nam caetera quid moror? omnes / Poenarum facies hic labor unus habet."

19. Jenny Saffran and Natasha Kirkham, "Infant Statistical Learning," *Annual Review of Psychology* 69 (2018): 181–203.

20. Virginia Woolf, "Craftsmanship," in *Collected Essays* (New York: Harcourt, 1953), 2:249.

21. Moth Radio Hour, http://exchange.prx.org/themoth.

22. Alexander Huth, Wendy de Heer, Thomas Griffiths, Frederic Theunissen, and Jack Gallant, "Natural Speech Reveals the Semantic Maps that Tile Human Cerebral Cortex," *Nature* 532, no. 7600 (2016): 453–458.

23. Nikolaus Kriegeskorte and Jörn Diedrichsen, "Peeling the Onion of Brain Representations," *Annual Review of Neuroscience* 42 (July 2019): 407–432; Mur, Bandettini, and Kriegeskorte, "Revealing Representational Content with Pattern-Information fMRI."

24. Robert Macfarlane, *Landmarks* (London: Penguin Random House UK, 2017).

25. Norman MacCaig, "By the Graveyard, Luskentyr," in *The Poems of Norman MacCaig*, ed. Ewen MacCaig (Edinburgh: Polygon, 2005), 431.

26. Alexander Huth, Shinji Nishimoto, An Vu, and Jack Gallant, "A Continuous Semantic Space Describes the Representation of Thousands of Object and Action Categories across the Human Brain," *Neuron* 76, no. 6 (2012): 1210–1224.

27. McFarlane, *Landmarks*.

28. William Joyce, *The Fantastic Flying Books of Mr. Morris Lessmore* (Cambridge, MA: Athenaeum Press, 2012).

29. Walt Whitman, "Song of the Open Road," in *Leaves of Grass* (New York: Doubleday and McClure Inc., 1926), 126.

30. John O'Keefe and J. Dostrovsky, "The Hippocampus as a Spatial Map: Preliminary Evidence from Unit Activity in the Freely-Moving Rat," *Brain Research* 34, no. 1 (1971): 171–175.

31. Torkel Hafting, Marianne Fyhn, Sturla Molden, May-Britt Moser, and Edvard Moser, "Microstructure of a Spatial Map in the Entorhinal Cortex," *Nature* 436, no. 7052 (2005): 801–806; Nathaniel Killian, Michael Jutras, and Elizabeth Buffalo, "A Map of Visual Space in the Primate Entorhinal Cortex," *Nature* 491, no. 7426 (2012): 761–764.

32. Timothy Behrens, Timothy Muller, James Whittington, Shirley Mark, Alon Baram, Kimberly Stachenfeld, and Zeb Kurth-Nelson, "What Is a Cognitive Map? Organizing Knowledge for Flexible Behavior," *Neuron* 100, no. 2 (2018): 490–509.

33. Edward H. Nieh, Manuel Schottdorf, Nicolas W. Freeman, Ryan J. Low, Sam Lewallen, Sue Ann Koay, Lucas Pinto, et al., "Geometry of Abstract Learned Knowledge in the Hippocampus," *Nature*, 2021.

34. Alexandra O. Constantinescu, Jill X. O'Reilly, and Timothy E. J. Behrens, "Organizing Conceptual Knowledge in Humans with a Gridlike Code," *Science* 352 (2016): 1464–1468.

35. Henri Poincaré, *Science and Hypothesis* (1902; repr., London: Walter Scott Publishing Co., 1914), xxiv.

36. Mons Garvert, Raymond Dolan, and Timothy Behrens, "A Map of Abstract Relational Knowledge in the Human Hippocampal-Entorhinal Cortex," *Elife* 6 (2017): e17086.

37. Nan Shepherd, *The Living Mountain* (Edinburgh: Canongate, 2019), 106.

38. Danielle S. Bassett and Marcelo Mattar, "A Network Neuroscience of Human Learning: Potential to Inform Quantitative Theories of Brain and Behavior," *Trends in Cognitive Sciences* 21, no. 4 (2017): 250–264.

39. Douglas Harper, *Online Etymological Dictionary*, 2001–2020, https://www.etymonline.com/.

40. Harang Ju and Danielle S. Bassett, "Dynamic Representations in Networked Neural Systems," *Nature Neuroscience* 23, no. 8 (2020): 908–917.

41. Jordan A. Litman and Charles D. Spielberger, "Measuring Epistemic Curiosity and Its Diversive and Specific Components," *Journal of Personality Assessment* 80, no. 1 (2003): 75–86.

42. Dominic Widdows, *Geometry and Meaning* (Stanford, CA: CSLI Publications, 2004).

43. Peter Gärdenfors, *Conceptual Spaces: The Geometry of Thought* (Cambridge, MA: MIT Press, 2004). Cf. Peter Gärdenfors, *The Geometry of Meaning: Semantics Based on Conceptual Spaces* (Cambridge, MA: MIT Press, 2014); Gaston Bachelard, *The Poetics of Space* (Paris: Presses Universitaires de France, 1958).

44. Susan Engel, *The Hungry Mind: The Origins of Curiosity in Childhood* (Cambridge, MA: Harvard University Press, 2015).

45. J. Michael McCarthy and Gim Song Soh, *Geometric Design of Linkages* (New York: Springer-Verlag, 2011); M. Muller, "A Novel Classification of Planar Four-Bar Linkages and Its Application to the Mechanical Analysis of Animal Systems," *Philosophical Transactions of the Royal Society B* 351, no. 1340 (1996): 689–720.

46. James Clerk Maxwell, "On Reciprocal Figures and Diagrams of Forces," *Philosophical Magazine* 4th Series 27 (1864): 250–261.

47. Perry Zurn and Danielle S. Bassett, "Network Architectures Supporting Learnability," *Philosophical Transactions of the Royal Society B* 375, no. 1796 (2020): 20190323.

48. Arianna Betti and Hein van den Berg, "Modelling the History of Ideas," *British Journal for the History of Philosophy* 22, no. 4 (2014): 812–835.

49. Arthur C. Benson, *Rossetti* (London, Macmillan and Co., Limited, 1904), 85–86.

50. Friedrich Nietzsche, *The Gay Science* (1882; repr., Cambridge: Cambridge University Press, 2001), §366. Nietzsche writes, "Our first questions about the value of a book, of a human being, or a musical composition are: Can they walk? Even more, can they dance?"

51. Jason Z. Kim, Zhixin Lu, Steven H. Strogatz, and Danielle S. Bassett, "Conformational Control of Mechanical Networks," *Nature Physics* 15 (2019): 714–720.

CHAPTER 8

1. Leanne Betasamosake Simpson, *As We Have Always Done: Indigenous Freedom through Radical Resistance* (Minneapolis: University of Minnesota Press, 2017), chapter 9.

2. Robin Wall Kimmerer, *Braiding Sweetgrass: Indigenous Wisdom, Scientific Knowledge and the Teachings of Plants* (Minneapolis: Milkweed Editions, 2013).

3. Kimmerer, *Braiding Sweetgrass*, 47; Simpson, *As We Have Always Done*, 231.

4. Susan Engel, *The Hungry Mind: The Origins of Curiosity in Childhood* (Cambridge, MA: Harvard University Press, 2015).

5. Perry Zurn and Arjun Shankar, "Introduction: What Is Curiosity Studies?," in *Curiosity Studies: A New Ecology of Knowledge*, ed. Perry Zurn and Arjun Shankar (Minneapolis: University of Minnesota Press, 2020), xi–xxx.

6. Naoki Higashida, *Fall Down 7 Times Get Up 8* (New York: Random House, 2017), 82.

7. Kristy Johnson, "Autism, Neurodiversity, and Curiosity," in *Curiosity Studies: A New Ecology of Knowledge*, ed. Perry Zurn and Arjun Shankar (Minneapolis: University of Minnesota Press, 2020), 129–146; Perry Zurn, "Cripping Curiosity: A Critical Disability Framework," *Curiosity and Power: The Politics of Inquiry* (Minneapolis: University of Minnesota Press, 2021), 149–171. Cf. Lydia X. Z. Brown, E. Ashkenazy, and Morenike Giwa, eds., *All the Weight of Our Dreams: On Living with Racialized Autism* (Lincoln, NE: DragonBee Press, 2017).

8. Naoki Higashida, *The Reason I Jump* (New York: Random House, 2016), 81.

9. Higashida, *Fall Down 7 Times Get Up 8*, 54, 147.

10. Higashida, *The Reason I Jump*, 59; Higashida, *Fall Down 7 Times Get Up 8*, 11.

11. Higashida, *The Reason I Jump*, 10, 36.

12. Higashida, *Fall Down 7 Times Get Up 8*, 43, 61, 156.

13. Higashida, *Fall Down 7 Times Get Up 8*, 184.

14. Higashida, *The Reason I Jump*, 139.

15. David Lydon-Staley, Dale Zhou, Ann Sizemore Blevins, Perry Zurn, and Danielle S. Bassett, "Hunters, Busybodies, and the Knowledge Network Building Associated with Deprivation Curiosity," *Nature Human Behavior* 5 (2020): 327–336.

16. Perry Zurn, *Curiosity and Power: The Politics of Inquiry* (Minneapolis: University of Minnesota Press, 2021), chapter 6.

17. Dale Zhou, David Lydon-Staley, Perry Zurn, and Danielle S. Bassett, "The Growth and Form of Knowledge Networks by Kinesthetic Curiosity," *Current Opinion in Behavioral Sciences* 35 (2020): 125–134.

18. Johnson, "Autism, Neurodiversity, and Curiosity."

19. Perry Zurn and Danielle S. Bassett, "On Curiosity: A Fundamental Aspect of Personality, a Practice of Network Growth," *Personality Neuroscience* 1, no. e13 (2018): 1–10.

20. David Lydon-Staley, Perry Zurn, and Danielle S. Bassett, "Within-Person Variability in Curiosity during Daily Life and Associations for Well-Being," *Journal of Personality* (2019): 1–17.

21. Perry Zurn and Danielle S. Bassett, "Network Architectures Supporting Learnability," *Philosophical Transactions B* 375, no. 1796 (2020); Zurn and Bassett, "On Curiosity."

22. David J. Connor, Jan W. Valle, and Chris Hale, eds., *Practicing Disability Studies in Education: Acting toward Social Change* (New York: Peter Lang, 2014).

23. Anne Meyer, David Howard Rose, and David Gordon, *Universal Design for Learning: Theory and Practice* (Wakefield, MA: CAST Professional Publishing, 2014); Elizabeth Berquist, *UDL: Moving from Exploration to Integration* (Wakefield, MA: CAST Professional Publishing, 2017).

24. Jan Doolittle Wilson, "Reimagining Disability and Inclusive Education through Universal Design for Learning," *Disability Studies Quarterly* 37, no. 2 (2017), https://dsq-sds.org/article/view/5417.

25. Bruce Bagemihl, *Biological Exuberance* (New York: St. Martin's Press, 2000).

26. Jimmy Garrett, "Freedom Schools (1965)," *Radical Teacher* 40 (1991): 42.

27. Liz Fusco, "Freedom Schools in Mississippi (1964)," *Radical Teacher* 40 (1991): 37.

28. Garrett, "Freedom Schools," 42.

29. Fusco, "Freedom Schools in Mississippi," 37.

30. "Mississippi Freedom School Curriculum," *Radical Teacher* 40 (1991): 8, 9, 6, 19.

31. Oskar Casquero, Ramon Ovelar, Jesus Romo, and Manuel Benito, "Reviewing the Differences in Size, Composition, and Structure between the Personal Networks of High- and Low-Performing Students," *British Journal of Educational Technology* 46, no. 1 (2015): 16–31; Manuel S. Gonzalez Canché and Cecilia Rios-Aguilar, "Critical Social Network Analysis in Community Colleges: Peer Effects and Credit Attainment," in *New Directions for Institutional Research* 163 (2015): 75–91; Mary K. Feeney and Margarita Bernal, "Women in STEM Networks: Who Seeks Advice and Support from Women Scientists," *Scientometrics* 85 (2010): 767–790; Jessica M. Dennis, Jean S. Phinney, and Lizette Ivy Chuateco, "The Role of Motivation, Parental Support, and Peer Support in the Academic Success of Ethnic Minority First-Generation College Students," *Journal of Student Development* 46, no. 3 (2005): 223–236; Inés Hernández-Avila, "Thoughts on Surviving as Native Scholars in the Academy," *American Indian Quarterly* 27, no. 1–2 (2003): 240–248; Allison Lombardi, Christopher Murray, and Jennifer Kowitt, "Social Support and Academic Success for College Students with Disabilities: Do Relationship Types Matter?," *Journal of Vocational Rehabilitation* 44, no. 1 (2016): 1–13; Brooke Erin Graham, "Queerly Unequal: LGBT+ Students and Mentoring in Higher Education," *Social Sciences* 8, no. 6 (2019): 1–19.

32. Alan J. Daly, ed., *Social Network Theory and Educational Change* (Cambridge, MA: Harvard Education Press, 2010).

33. Paulo Freire, *Pedagogy of the Freedom: Ethics, Democracy, and Civic Courage* (Lanham, MD: Rowman and Littlefield, 1998), 32, 36, 41.

34. bell hooks, *Teaching to Transgress: Education as the Practice of Freedom* (New York: Routledge, 1994), 2, 8, 60.

35. Eli Meyerhoff, *Beyond Education: Radical Studying for Another World* (Minneapolis: University of Minnesota Press, 2019).

36. Britta Renner, "Curiosity about People: The Development of a Social Curiosity Measure in Adults," *Journal of Personality Assessment* 87, no. 3 (2006): 305–316; Yu-Hui Fang, Kwei Tang, Chia-Ying Li, and Chia-Chi Wu, "On Electronic Word-of-Mouth Diffusion in Social Networks: Curiosity and Influence," *International Journal of Advertising* 37, no. 3 (2018): 360–384; Engel, *The Hungry Mind*, 128–147.

37. Mario Livio, *Why? What Makes Us Curious* (New York: Simon and Schuster, 2017), 10.

38. Sigmund Freud, *Leonardo da Vinci and a Memory of His Childhood* (New York: W. W. Norton and Company, 1910), 25.

39. Kimmerer, *Braiding Sweetgrass*, 40.

40. Alison Kafer, *Feminist Queer Crip* (Bloomington: Indiana University Press, 2013), 2.

41. Gloria Anzaldúa, "Interview with Gloria Anzaldúa," in *Borderlands / La Frontera* (San Francisco: Aunt Lute Books, 1987), 230.

42. Shay Welch, *The Phenomenology of a Performative Knowledge System: Dancing with Native American Epistemology* (New York: Palgrave, 2019), 17.

43. Zurn, *Curiosity and Power*.

44. Zurn and Shankar, "Introduction: What Is Curiosity Studies?"

45. Rosemary E. Ommer, "Curiosity, Interdisciplinarity, and Giving Back," *ICES Journal of Marine Science* 75, no. 5 (2018): 1526–1535; Willard McCarty, "Becoming Interdisciplinary," *A New Companion to Digital Humanities*, ed. Susan Schreibman, Ray Siemens, and John Unsworth (Hoboken, NJ: Wiley-Blackwell, 2016), 69–83.

46. Tanya Augsburg, "Becoming Transdisciplinary: The Emergence of the Transdisciplinary Individual," *World Futures: The Journal of Global Education* 70, no. 3 (2014): 233–247; Basarab Nicolescu, *Manifesto of Transdisciplinarity* (Albany: SUNY Press, 2002).

47. Andrew E. Pelling, "Re-Purposing Life in an Anti-Disciplinary and Curiosity-Driven Context," *Leonardo* 48, no. 3 (2015): 274–275; Petar Jandric, "The Methodological Challenge of Networked Learning: (Post)disciplinarity and Critical Emancipation," in *Research, Boundaries, and Policy in Networked Learning*, ed. Thomas Ryberg, Christine Sinclair, Sian Bayne, and Maarten de Laat (Cham, Switzerland: Springer, 2016), 165–181.

48. Miranda Fricker, *Epistemic Injustice: Power and the Ethics of Knowing* (Oxford: Oxford University Press, 2007).

49. Jonathan Paul Eburne, *Outsider Theory: Intellectual Histories of Unorthodox Ideas* (Minneapolis: University of Minnesota Press, 2018).

50. Brian Burkhart, *Indigenizing Philosophy through the Land: A Trickster Methodology for Decolonizing Environmental Ethics and Indigenous Futures* (Lansing: Michigan State University Press, 2019).

CONCLUSION

1. Anna Dufourmantelle, *In Praise of Risk* (New York: Fordham University Press, 2019), 170.

2. Charles Darwin, *The Power of Movement in Plants* (1880; repr., New York: D. Appleton and Company, 1896), 573.

3. Elias Lönnrot, *The Kalevala: The Epic Poem of Finland* (1887; repr., Woodstock, ON: Devoted Publishing, 2016), 16.

4. Virginia Woolf, *Collected Essays* (New York: Harcourt, Brace, and World, 1925), 2:250–251, 278.

5. Gloria Anzaldúa, *Light in the Dark / Luz en lo Oscuro* (Durham, NC: Duke University Press, 2015), 92, 84, citing Leonard Cohen, "Anthem" (1992).

6. Friedrich Nietzsche, *Thus Spoke Zarathustra* (1892; repr., New York: Penguin Books, 1966), 189.

7. Friedrich Nietzsche, *Beyond Good and Evil* (1886; repr., New York: Vintage Books, 1989), 9.

8. Friedrich Nietzsche, *Human All Too Human* (1886; repr., Cambridge: Cambridge University Press, 1986), preface §1.

9. Maurice Merleau-Ponty, *The Visible and the Invisible* (Evanston, IL: Northwestern University Press, 1968), 148–149.

10. Katharine T. Bartlett, "Cracking Foundations as Feminist Method," *Journal of Gender, Social Policy and the Law* 8, no. 1 (2000): 31–54.

11. Sara Ahmed, *Willful Subjects* (Durham, NC: Duke University Press, 2014).

12. Fred Moten, *In the Break: The Aesthetics of the Black Radical Tradition* (Minneapolis: University of Minnesota Press, 2003), 85, 91, 24.

APPENDIX

1. Apuleius, *Metamorphoses* (Cambridge, MA: Harvard University Press, 1989).

2. Friedrich Nietzsche, *On the Genealogy of Morals*, trans. Walter Kaufmann and R. J. Hollingdale (New York: Vintage Books, 1989), preface §1.

3. Margaret Cavendish, "Similizing the Head of Man to a Hive of Bees," in *Margaret Cavendish's Poems and Fancies: A Digital Critical Edition*, ed. Liza Blake, May 2019, http://library2.utm.utoronto.ca/poemsandfancies/.

4. Nicholas A. Basbanes, *A Gentle Madness* (New York: Henry Holt, 1995), 31.

5. Virginia Woolf, "Madame de Sévigné," in *Collected Essays* (New York: Harcourt, 1953), 3:66–67.

6. Emily Rose, *Translating Trans Identity: (Re)writing Undecidable Texts and Bodies* (New York: Routledge, 2021), 105.

7. Ruth Stiles Gannett, *Elmer and the Dragon* (New York: Random House, 1950), 41–44.

8. Alastair Reed, "Curiosity," in *Weathering: Poems and Translations* (Edinburgh: Canongate, 1978), 53.

9. Jean Craighead George and John George, *The Dipper of Copper Creek* (Boston: E. P. Dutton and Co., 1967), 20, https://archive.org/stream/dipperofcoppercr00ingeor/dipperofcopper cr00ingeor_djvu.txt.

10. Norton Juster, *The Phantom Tollbooth* (New York: Random House, 2001), 169–172.

11. Craighead George and George, *The Dipper of Copper Creek*, 21.

12. Saint Augustine, *The Confessions*, trans. William Watts (Cambridge, MA: Loeb Classical Library, 2000), X.35.

13. Plato, *Apology*, in *Euthyphro, Apology, Crito, Phaedo, and Phaedrus* (Cambridge, MA: Loeb Classical Library, 1904), 30c–31c.

14. Virginia Woolf, "Why?," in *Collected Essays* (New York: Harcourt, 1925), 2:280.

15. Cesare Ripa, *Iconoglia, or Moral Emblems* (1603), trans. P. Tempest (London: Benjamin Motte, 1709).

16. Arnold Stark Lobel, *Grasshopper on the Road* (New York: HarperCollins Publishers, 1978), 46–52.

17. Archilochus, frag. 201, cited in Martin Litchfield West, ed., *Iambi et Elegi Graeci*, vol. 1 (Oxford: Oxford University Press, 1971).

18. Beatrix Potter, *The Tale of Mr. Tod* (London: Frederick Warne, 1905); Beatrix Potter, *The Tale of Mrs. Tiggy-Winkle* (London: Frederick Warne, 1905).

19. Plutarch, "On Being a Busybody," *Moralia VI*, trans. W. C. Helmbold (Cambridge, MA: Loeb Classical Library, 2005), 520e.

20. Legend of Tu-Tok-A-Nu'-La (El Capitan): A Miwok Legend, accessed April 6, 2021, https://www.firstpeople.us/FP-Html-Legends/LegendofTu-Tok-A-Nu-La-Miwok.html.

21. Julie Parker, *The Story of Tu-Tok-A-Nu-La*, accessed April 6, 2021, http://www.alpinist .com/doc/web19s/wfeature-story-of-tu-tok-a-nu-la.

22. Legend of Tu-Tok-A-Nu'-La (El Capitan).

23. H. A. Rey and Margret Rey, *Curious George* (Boston: HMH Books for Young Readers, 2016).

24. Maya Zhe Wang and Benjamin Y. Hayden, "Monkeys Are Curious about Counterfactual Outcomes," *Cognition* 189 (2019): 1–10.

25. Ian Leslie, *Curious* (New York: Basic Books, 2014), 149. Indeed, Antonelli's Twitter handle is @curiousoctopus.

26. Lewis Carroll, *Alice's Adventures in Wonderland and through the Looking-Glass* (New York: Modern Library, 2002), 4.

27. Cael Keegan, *Lana and Lilly Wachowski: Sensing Transgender* (Champaign: University of Illinois Press, 2018).

28. Frances Hodgson Burnett, *The Secret Garden* (London: Folio Society, 2006), 34.

29. Tim Ingold, *The Life of Lines* (New York: Routledge, 2015), 47.

30. Burnett, *The Secret Garden*, 35.

31. Genesis 3; Saint Augustine, *On Genesis*, trans. Roland J. Teske (Washington, DC: Catholic University Press, 1991); Thomas Aquinas, *Summa Theologica* (1274; repr., New York: Benziger Brothers, Inc., 1947–1948).

32. Gloria Anzaldúa, *Light in the Dark / Luz en lo Oscuro* (Durham, NC: Duke University Press, 2015), 121.

33. Maria Sibylla Merian, *Metamorphosis insectorum Surinamensium 1705*, ed. Marieke van Delft and Hans Mulder, trans. Patrick Lennon (Amsterdam: Lanoo Publishers, 2001), 179, plate 10.

Index

Page numbers in italics refer to figures.